# FASHIONOPOLIS

Also by Dana Thomas

*Gods and Kings*

*Deluxe*

# FASHIONOPOLIS

## THE PRICE OF FAST
## FASHION & THE FUTURE
## OF CLOTHES

## DANA THOMAS

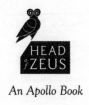

*An Apollo Book*

First published in the US in 2019 by Penguin Press,
an imprint of Penguin Random house LLC

First published in the UK in 2019 by Apollo,
an imprint of Head of Zeus Ltd

Photo credits on page 295

9 7 5 3 1 2 4 6 8

A catalogue record for this book is available from
the British Library.

ISBN (HB): 9781789546064
ISBN (XTPB): 9781789546071
ISBN (E): 9781789546057

Typeset by Claire Vaccaro

Printed and bound in Great Britain by
CPI Group (UK) Ltd, Croydon, CR0 4YY

Head of Zeus Ltd
First Floor East
5–8 Hardwick St
London EC1R 4RG
WWW.HEADOFZEUS.COM

To Hervé

*and*

our light,

Lucie Lee

But seest thou not
what a deformed thief this fashion is?

William Shakespeare, *Much Ado About Nothing*, Act 3, Scene 3

# Contents

# Introduction

WHEN AMERICAN First Lady Melania Trump traveled to visit migrant children in a Texas detention center in 2018, she was enrobed in an olive-drab anorak by the Spanish fast-fashion retailer Zara, with these words scrawled, graffiti-like, in white, on the back:

I REALLY DON'T CARE, DO U?

Pundits opined that the jacket broadcasted how Mrs. Trump truly felt about the locked-up kids. Or her public duties. Or her marriage. Her husband tweeted that it was her view of "the Fake News Media." Her spokeswoman claimed: "There was no hidden message."

She was right, in a sense. The message was loud and clear. And it's a devastating reflection of how we live now.

The jacket was, in effect, the most existential garment ever designed, made, sold, and worn.

Zara is the world's largest fashion brand. In 2018, it produced more than 450 million items. Its parent company,

Spain-based Inditex, reported €25.34 billion, or $28.63 billion, in sales for 2017, of which Zara made up two-thirds, or approximately $18.8 billion.

The jacket, which came from the company's Spring-Summer 2016 collection, retailed for $39. To be able to sell clothing that cheaply and still reap a sizable profit, production is outsourced to independently owned factories in developing nations, where there is little or no safety and labor oversight and wages are generally poverty level, or lower.

At the time workers were cutting and sewing Mrs. Trump's jacket, Amancio Ortega, the octogenarian cofounder and former chairman of Inditex, was the second-richest person in the world (after Bill Gates), with a net worth of $67 billion.

The jacket itself was made of cotton. Conventionally grown cotton is one of agriculture's most polluting crops. Almost one kilogram (2.2 pounds) of hazardous pesticides is required to grow one hectare—or two and a half acres—of the fluff.

It was dyed and lettered with coloring agents that, while decomposing in landfill, would poison the earth and groundwater.

On average—*average*—the piece would be worn seven times before getting tossed. Though given the criticism hurled at Mrs. Trump for donning it on that visit, she would likely never put it on again. So, like most clothing today, to the dump the jacket would go.

*"I really don't care, do you?"*


EACH DAY, we wake up and pose an elemental question:

"What am I going to wear?"

Much thought goes into the decision: *How do I feel? What's the weather? What do I have to do? What do I want to say? To project?*

Clothes are our initial and most basic tool of communication. They convey our social and economic status, our occupation, our ambition, our self-worth. They can empower us, imbue us with sensuality. They can reveal our respect, or our disregard, for convention. "Vain trifles as they seem," Virginia Woolf wrote in *Orlando*, "clothes . . . change our view of the world and the world's view of us."

As I sit here and write this, I'm wearing a black cotton jersey dress with a white pointed collar and shirt cuffs, made in Bangladesh. I spotted it on a Facebook ad, clicked through, and within days it was delivered to my home. It is flattering and fashionably on point. But did I think hard about where it came from when I ordered it? Did I consider why it only set me back thirty bucks? Did I need this dress?

No. No. And nope.

I am not alone.

Every day, billions of people buy clothes with nary a thought—nor even a twinge of remorse—about the consequences of those purchases. In 2013, the Center for Media Research declared that shopping was becoming "America's favorite pastime." Shoppers snap up five times more clothing now than they did in 1980. In 2018, that averaged sixty-eight garments a year. As a whole, the world's citizens acquire 80 billion apparel items annually.

And if the global population swells to 8.5 billion by 2030, and GDP per capita rises by 2 percent in developed nations and 4 percent in developing economies each of those intervening years, as experts predict, and we don't change our consumption habits, we will buy 63 percent *more* fashion—from 62 million tons to 102 million tons. This is an amount, the Boston Consulting Group and the Global Fashion Agenda report, that would be the "equivalent of 500 billion T-shirts."

All this is by design. In airports, you can pick up an entire new wardrobe on the way to the gate. In Tokyo, you can score a tailored suit from a vending machine. Love that outfit on Instagram? *Click-click*, and it's yours. Walk into a fashion store: techno thumps; surfaces gleam; the light is desert-sharp— ah, the better to see the abundance of offerings. A freneticism sets in. Price, curiously, becomes moot. You're so beguiled, and so overstimulated, you forget to consider such fundamentals as quality. "It's like a sex shop," a former fashion magazine editor mused as we discussed it over lunch in Paris one day. "Or a Vegas casino," I countered. You spend freely, recklessly even, and though you've probably been rooked, you feel like you've won.

"The expectation is to keep up with the ever-changing trends—[to] respond to the constant noise that says, 'Come buy something else,'" Dilys Williams, director of the Centre for Sustainable Fashion at the London College of Fashion, told me. "The original, pre-industrial definition of fashion was to make things together—a collective that is a convivial, sociable process we use to communicate with each other. The current definition is the production, marketing, and consumption of clothes—an industrialized system for making money."

And it's not sustainable. None of it.

SINCE THE INVENTION of the mechanical loom nearly two and a half centuries ago, fashion has been a dirty, unscrupulous business that has exploited humans and Earth alike to harvest bountiful profits. Slavery, child labor, and prison labor have all been integral parts of the supply chain at one time or another—including today. On occasion, society righted the wrongs, through legislation or labor union pressure. But trade deals, globalization, and greed have undercut those good works.

Up until the late 1970s, the United States produced at least 70 percent of the apparel that Americans purchased. And—thanks to the New Deal—for much of the twentieth century, brands and manufacturers were expected to adhere to strict national labor laws. But in the late 1980s, a new segment of the apparel business cropped up: "fast fashion," the production of trendy, inexpensive garments in vast amounts at lightning speed in subcontracted factories, to be hawked in thousands of chain stores. To keep the prices low, fast-fashion brands slashed manufacturing costs—and the cheapest labor was available in the world's poorest countries. Offshoring caught on across the industry, just as globalization was unfurling. Though it started as a small corner of the business, fast fashion's astounding success was so enviable it soon reset the rhythm for how clothing—from luxury to athletic wear—was and is conceived, advertised, and sold. The impact was dramatic: in the last thirty years, fashion has grown from a $500 billion trade, primarily domestically produced, to a $2.4-trillion-a-year global behemoth.

The fallout has been great.

The first hit was to labor in developed economies. In 1991, 56.2 percent of all clothes purchased in the United States were American-made. By 2012, it was down to 2.5 percent. According to the Bureau of Labor Statistics, between 1990 and 2012, the US textile and garment industry lost 1.2 million jobs. That was more than three-fourths of the sector's labor force, siphoned to Latin America and Asia. Once-vibrant industrial centers down the Eastern Seaboard and across the South faded into ghost towns, as factories sat empty and those who were laid off went on unemployment. In the United Kingdom in the 1980s, one million worked in the UK textile industry; now, only one hundred thousand do. The same went down across most of western Europe. All while apparel and textile jobs globally nearly doubled, from 34.2 million to 57.8 million.

Offshoring created massive and crippling trade deficits in the West. In 2017, US apparel exports totaled roughly $5.7 billion, while imports were about $82.6 billion. In 2017, Britain imported 92.4 percent of its clothing. In the EU, only Italy managed to hold tight, since the "Made in Italy" label implies quality and confers cachet in the luxury fashion market.

Sometimes an offshoring scandal would make the news. In the summer of 2012, Ralph Lauren came under fire for having the uniforms he designed for the US Olympic team made in China. *Forbes* called it "clearly a PR disaster." Senate Majority Leader Harry Reid, a Democrat from Nevada, said that the US should "burn" them. Speaker of the House of Representatives John Boehner, a Republican from Ohio, charged that Ralph Lauren and his executives "should have known better."

But what upset the pols didn't fluster consumers—quite the contrary. They understood, as Lauren did, that cost trumps all other concerns.

According to a 2016 poll, when given the choice between buying a $50 pair of pants made offshore or an $85 pair manufactured in the US, 67 percent of respondents said they'd go for the cheaper ones. The response was the same even if their annual household income was more than $100,000.

The fast-fashion revolution has been grossly lucrative for the entire industry. In 2018, five of the world's fifty-five richest individuals were fashion company owners. Not counting the three Waltons of Walmart.

THE SECOND CASUALTY of the age of fast fashion has been human rights in developing nations. Fashion employs one out of six people on the globe, making it the most labor-intensive industry out there—more than agriculture, more than defense. Fewer than 2 percent of them earn a living wage.

Most apparel workers are women; some are boys and girls. In 2016, H&M, Next, and Esprit were found to have Syrian refugee children sewing and hauling bundles of clothes in subcontracted workshops in Turkey. (The brands have reportedly since rectified the situation.) Some factories are so shoddy they catch fire, or worse, collapse. Because pay is egregiously low, workers are forced to find less reputable ways to make ends meet.

"In Sri Lanka, we met a female worker who had a toothache. She had to take a loan to pay for it because, on her wage, she couldn't afford a dentist appointment," an NGO official told a standing-room crowd at SOCAP17, a conference in San Francisco "dedicated to accelerating a new global market at the intersection of money and meaning."

"She couldn't afford to pay back the loan," the advocate continued, "so she had to become a sex worker to make the money to pay it off. All while still making clothes that you and I wear for a very well-known and big supplier."

The third victim has been Earth. Fashion's speed and greed has eviscerated the environment in all ways. The World Bank estimates that the sector is responsible for nearly 20 percent of *all* industrial water pollution annually. It releases 10 percent of the carbon emissions in our air; 1 kilogram of cloth generates 23 kilograms of greenhouse gases.

The fashion industry devours one-fourth of chemicals produced worldwide. The creation of one cotton T-shirt requires a third of a pound of lab-concocted fertilizers and 25.3 kilowatts of electricity, and the World Wildlife Fund (WWF) has stated it can take up to 2,700 liters of water to grow the cotton.

Synthetic fabrics release microfibers into water when washed, both at mills and at home. Up to 40 percent enter rivers, lakes, and oceans; are ingested by fish and mollusks; and worm their way up the food chain to humans, researchers at the University of California in Santa Barbara reported in 2016. That same year, nearly 90 percent of 2,000 fresh- and seawater samples tested by the Global Microplastics Initiative contained microfibers. In 2017, Greenpeace found microfibers in the waters of Antarctica.

Of the more than 100 billion items of clothing produced each year, 20 percent go unsold—the detritus of "economies of scale." Leftovers are usually buried, shredded, or incinerated, as Burberry embarrassingly admitted in 2018.

In the last twenty years, the volume of clothes Americans throw away has doubled—from 7 million to 14 million tons. That equals 80 pounds per person per year. The European Union disposes of 5.8 million tons of apparel and textiles a year. Worldwide, we jettison 2.1 *billion* tons of fashion. Much of it is shunted to Africa, our rationalization being that the poorest continent needs free clothing. In 2017, USAID reported that the East African Community (EAC), an association comprised of Kenya, Uganda, Tanzania, Burundi, Rwanda, and South Sudan, imports as much as $274 million worth of used clothes each year. Kenya alone accepts 100,000 tonnes annually. Some of these used togs are resold by secondhand merchants at a steep discount—a pair of jeans, for example, will run $1.50 in Nairobi's Gikomba Market. Our fashion bulimia has so decimated the continent's indigenous apparel business that, in 2016, the EAC adopted a three-year phase out of the importation of hand-me-downs. In response, in 2018, the Trump administration threatened to launch a trade war—stating that the ban would lead to the loss of 40,000 jobs in the US—and the EAC backed down, with the exception of Rwanda; the administration continued to menace the small country.

And the rest of our leftovers?

Landfill.

The Environmental Protection Agency reported that Americans sent 10.5 tons of textiles, the majority of which were clothes, to landfill in 2015. (The EPA during the Trump administration has not released an updated figure.) In the UK, 9,513 garments are dumped every five minutes; textiles are the country's fastest-growing waste stream. Most clothing contains synthetics, and most synthetics are not biodegradable. As with Mrs. Trump's Zara jacket, the fabrics that do break down often contain chemicals that contaminate soil and the water table.

Some brands have pushed back. In 2011, the pro-environment American outdoor gear company Patagonia took out a full-page advertisement in the *New York Times* on Black Friday—the day after Thanksgiving, and traditionally the kickoff of the Christmas shopping season—that featured a photograph of a zip-up fleece and the copy line: "Don't Buy This Jacket." The ad confessed that the production of the shell "required 135 liters of water, enough to meet the daily needs (three glasses a day) of forty-five people," "generated nearly twenty pounds of carbon dioxide," and left behind "two-thirds its weight in waste . . . This jacket comes with an environmental cost higher than its price." (And this was before the discovery of microfibers in our waterways.) "We ask you to buy less and to reflect before you spend a dime on this jacket or anything else."

The act of buying and running the ad made news around the world. But its actual message fell on deaf ears. The National Retail Federation reported

Americans plunked down a record-breaking $52.4 billion in those four days, a 16 percent increase over the 2010 total of $45 billion.

*"I really don't care, do you?"*

*POLIS* IN ANCIENT GREEK meant "city." The Greek philosopher Plato put forth in the Socratic dialogue *The Republic* that an ideal polis should embody four cardinal virtues: wisdom, courage, moderation, and justice. If all came together harmoniously, the polis would attain perfect equality—a "just city."

The English city of Manchester in the eighteenth century was the birthplace of the Industrial Revolution and the apparel system as we know it today. Renowned for its tremendous scale of production, "Cottonopolis," as it was christened, was the world's first major manufacturing center, captained by tycoons, who essentially enslaved battalions of workers.

A hundred years on, the German expressionist filmmaker Fritz Lang illustrated the perniciousness of such social and economic unbalance in his silent picture *Metropolis*. The sci-fi epic forecast a dystopian future where the underclass moils in grim subterranean factories for the financial benefit of a happy few in shining skyscrapers. Our technology has evolved; our ethos has not.

In their own ages, Cottonopolis and *Metropolis* embodied capitalism with no motive other than profits. In today's Fashionopolis, we have Manchester and Lang on a global scale.

The history of the rag trade is dark, but not completely so. There was a midcentury moment when the garment industry did some things right—when people knew those who cut and sewed their clothes. They went to the same church. Or their kids attended school together. Or they were related. There were injustices, to be sure. But not to the degree of today; because of proximity, consumers couldn't turn a blind eye. That is no longer the case.

We imagine ourselves as more learned, more egalitarian, more humane than our predecessors. More *woke*. That by procuring $5 tees and $20 jeans by the sackful, we aren't causing grievous harm. We might even be creating good jobs on the other side of the world for those in need. Having visited many offshore factories and spoken with dozens of workers, I can assure you this is not reality.

But during my reporting, I have also found many reasons to remain hopeful. Through the Herculean efforts of brave advocates, creators, entrepreneurs, innovators, investors, and retailers, and the unfeigned demands of a rising generation of conscientious consumers, the apparel industry is being forced to veer toward a more principled value system.

Visionaries throughout the world are recasting the business model with hyperlocalism in rural areas like the American South; a return of (smarter) manufacturing in New York, Los Angeles, and across Europe; a cleaner denim process from cotton fields to finishing plants; a holistic approach to luxury that will trickle down from the Paris runway to the online resellers; scientific breakthroughs that are creating truly circular fabrics; technological advances that will completely change how apparel is made; and the total and rapid rethinking of how we buy what we wear.

More than a decade ago, the slow food and organic movement prodded us to be more informed about what we eat and to contemplate the consequences of alimentary industrialization. The same has not happened broadly with fashion. Yet.

As it was with the sustainable food crusade, fashion's changemakers are striving to bring sourcing and producing back to a human scale, in a modern and mediated way. Many are working toward a vertically integrated system, to keep the entire process in one location and avoid the troubles that come with a global, opaque supply chain. The Fashionopolis of the future could be good, possibly even just.

We, as consumers, play a pivotal part. It's time to quit the mindless shopping and consider what we are doing, culturally and spiritually. To fan the change, we need to understand how we got to this point.

We have to look at the game that is Fashionopolis.

Only then can we make something better.

When we ask ourselves, "What am I going to wear today?" we should be able to answer knowledgeably and with a dash of pride.

We have been casual about our clothes, but we can get dressed with intention.

It is time to really care.

# part one

# ONE

# Ready to Wear

ON THE THIRD NIGHT of the seventy-first Cannes Film Festival, in May 2018, Australian actress and jury president Cate Blanchett floated down the red carpet in a showstopping sleeveless bubble-hem gown. The rowdy floral print began on the bodice as black-and-white line drawings, like paint-by-numbers, and eventually erupted on the voluminous skirt into full-blown Technicolor. Daring and complicated, it was exquisitely executed by Mary Katrantzou, a Greek-born, London-based women's wear designer most consumers have never heard of, though have most likely worn, after a fashion.

Katrantzou is one of the talents who creatively fuels the Fashionopolis machine: the original silhouettes she dreams up in her small London atelier are sold in limited numbers by luxury retailers in cosmopolitan capitals. This is the apex of the fashion pyramid—the same place where Gucci creative director Alessandro Michele, Louis Vuitton menswear designer Virgil Abloh, Givenchy artistic director Clare Waight Keller, and other high fashion designers sit. The clothes that

Katrantzou and her confreres design are copied by fast-fashion brands on the cheap and peddled in chain stores—the massification that constitutes the broad bottom of the fashion pyramid. The "knockoffs," as the fakes are known, rake in millions for the hucksters who vend them. Katrantzou, meanwhile, receives nothing from the unauthorized global rollout of her work: no money, no glory, no acknowledgment that she ignites trends or contributes to the fashion conversation. She toils; others profit; we all wear.

Sound unfair? It is. But this trickle-down scheme—as sharply laid out by Meryl Streep in the "cerulean-blue sweater" scene in the film *The Devil Wears Prada*—is how the fashion industry works.

It begins simply enough at a wholly unglamorous semiannual trade show outside of Paris, near Charles de Gaulle airport, called Première Vision Paris. For three days each February and September, more than sixty thousand apparel trade professionals from 120 countries descend on the multihall convention complex of Villepinte to shop the world's largest selection of fabrics and textile designs, leather, accessories, and manufacturing innovations in one place—for the February 2019 edition, there were 1,900 exhibitors. One hall is dedicated to yarns, fabrics, and sourcing solutions. Another to design, and more fabrics—some 20,000 in all. Another to leather—there are 10,000 of those. Another to accessories. The endless rows of office-gray sales stands are punctuated with installations highlighting the season's tendencies as set by thread suppliers, color companies such as Pantone, and an army of consultants who specialize in trend forecasting. Première Vision is where every major—and many a minor—fashion brand's new season begins to take shape.

In the winter of 2018, I accompanied Katrantzou's fabric expert Raffaella Mandriota, a twenty-seven-year-old Italian metalhead who favors Maison Margiela Tabi boots, on her two-day expedition to Première Vision—or "PV" in fashion-speak. She was scouting for the Spring-Summer 2019 women's wear collection, to be presented on a London runway nine months hence.

Her first stop was the top Italian mill, Canepa—one of her regular suppliers. After a quick hello, and an espresso, she whipped through racks and racks of jacquards, prints, and solids, called "bases," giving each one a proper look, if only for a tenth of a second, and a feel, to understand texture and pliability. When she thought one might work, she pulled it and placed it on her mounting pile—or "selection"—on the table. When she finished— in ten minutes, max—a company rep wrote up the order.

Mandriota does this twelve to fifteen times in a day at PV, thus the need for coffee; her to-do list was *long*. There were definite trends on view— natural dyes, seersucker, Candy Land–like colors, Lurex, hemp, iridescent silks—but the range that Mandriota chose was broad: an orange devoré on black chiffon; a kelly-green viscose; a white polyester waffle-weave; a gray silk inkjet-printed with a cloudy sky; a navy, black, and evergreen Fortuny-style pleated silk with green and blue coral motif. "Beautiful, this one," she said, as she set the coral swatch on her stack. "Mary *loves* pleats."

Mandriota watched her budget carefully. "Chinese silk has gotten so expensive," she told me. "A twenty percent increase just this season. The Chinese have increased domestic consumption, so they export less. And because of pollution, the silkworm cocoons are dying."

Throughout her hunt, Mandriota peppered her vendors with questions: "What's the minimum order?" "Anything organic or sustainable?" "What other colorways are available?" "Can you print on this?" "Can you emboss on wool?" "Could you interpret Mary's design with this same jacquard technique?" In our two ten-hour days at Première Vision, she ordered at least a thousand samples.

Six weeks later, cartons filled with swatches began to arrive at Katrantzou's studio, a prewar loft in Islington. She and her assistants conducted a first edit, and a second, and on and on, until they had whittled the mass down to a manageable array that could tell the season's story.

A Hellenistic beauty, with mink-like eyes and hair to match that falls straight to her elbows, Katrantzou was born in Athens in 1983 to a retailing family: her grandfather founded Katrantzou Spor, which was the city's largest department store until it was burned to the ground during Greece's political riots of the 1970s. Her father worked in security, and her mother had an interior design shop and furniture factory.

In 2003, Katrantzou went to the United States to study interior architecture at the Rhode Island School of Design in Providence. Halfway through her sophomore year, she moved to London as an exchange student at Central Saint Martins College of Art and Design to learn textile design for interiors. "I loved the idea that textiles were about surface," she told me. "And there was an immediacy about it that I hadn't found in architecture."

Hooked, she stayed and earned a bachelor's degree in textile design and a master's in fashion with a focus on prints—at a time when fashion printmaking was moving from the more artisanal silk screen process, in which a piece of mesh cloth (originally silk), etched with an image, is stretched over a wood frame and squeegeed with ink, to digital, which is computer drawn and generated. For her MA degree show, in February 2008, she sent out ten identically shaped dresses printed with a trompe l'oeil of gigantic jewelry—magnifying common items in print on cloth has since become her leitmotif. With a grant from the British Fashion Council's young talent fund, NEWGEN, she launched her own brand during London Fashion Week the following September and landed several influential retailers, including Browns in London, Joyce in Hong Kong, and Colette in Paris.

In 2011, she won the British Fashion Award for Emerging Talent in women's wear. She has lived up to her promise—so much so that in early 2018, she sold a stake of her firm to Hong Kong–based Yu Holdings, a start-up fund founded and run by the ambitious twenty-seven-year-old Chinese fashion and technology investor Wendy Yu. (Weeks later, Yu also announced her firm was endowing the position of curator in charge at the Metropolitan Museum of Art's Costume Institute.) For Katrantzou, Yu earmarked a $20 million infusion—money that will encourage brand growth. "I think Mary could be a global lifestyle brand in the next ten to twenty years," Yu said.

As the new collection would celebrate Katrantzou's tenth anniversary, she decided it should be a best-of—meaning she'd rework former print patterns and silhouettes in a more modern, and mature, way. For themes, she chose blown-glass perfume bottles; vintage postage stamps; nature, such as insects, butterflies, and seashells; and the arts. Mandriota asked some of the textile suppliers to redo jacquard samples she'd selected at PV in Katrantzou's new print designs.

In early May, Katrantzou sat down at a wood IKEA table with Mandriota and women's wear head Gregory Amore to review the new iterations. One jacquard, a quilt-like brocade from the Italian mill Ostinelli Seta, had been proposed as chrysanthemums in grainy blues. Katrantzou kept the fabric and the technique, but replaced the flowers with a collage she created: mounds of gems, baubles, and ropes of pearls, like you'd find in a pirate's treasure chest, scattered across a coral seabed, in a palette of burnt orange, Mediterranean azure, antique gold, and iridescent white.

For solids, Katrantzou had snippets of stretch cotton poplin, sourced from various companies, that had been custom dyed, or "lab dipped," in Pantone hues she chose—sandy beige, Tiffany blue, sunshine yellow—to see how the different offerings took the color. She wasn't thrilled with the first test: the tones were a tad dull, as if rinsed in murky dishwater. Another—from a different supplier—she warmed to immediately. The colors were truer, and the quality of the cloth was clearly superior.

"Everything about this feels lighter, and somehow thicker," she said as she caressed one of the swatches. "It feels very substantial."

"It's more precious," Mandriota said.

"But it is double the price," Katrantzou said.

"Yes."

When final fabric choices arrived from mills, Katrantzou pulled a few panels and sent them off to Mumbai to be embroidered. (As needlework is still a valued skill in India, it is a center for handcrafted fashion embellishments.) Garment samples were made, either by her in-house atelier or a contracted factory—she relies on two in Italy, one in Portugal, and three small family-owned workshops in the UK that do short runs of twenty to fifty pieces. For six weeks, she conducted fittings on her longtime house model Julia, a leggy blonde from Sweden. "Julia has an opinion on the aesthetic," Katrantzou said. "She understands the fit, and advises us."

In late June, Katrantzou presented her finished "precollection," which is in the more commercial range, to retailers in a Paris showroom she rented during menswear week. (Buyers hit men's shows and women's precollections on the same trip.) The space was gloriously belle epoque Parisian: a one-floor walk-up with patinaed oak paneling, herringbone parquet, and arched windows with a view on the Place des Victoires. The finished clothes, hanging on racks and presented informally on models, were colorful and enticing. Retail buyers examined each item thoughtfully and, over coffee and petit fours at small tables, placed their orders. Katrantzou sat down with them and listened to their opinions and observations. Sometimes she'll tweak her designs to incorporate their counsel.

The flashiest, most photographable pieces—the sort like Blanchett wore in Cannes—Katrantzou saved for her "show collection," which she presented during London Fashion Week in September. At eight on a Saturday night, retailers, editors, bloggers, and scribes filed into the Roundhouse, a rock concert venue in north London, for the show. Thirty-five models slowly emerged, one by one, and walked the circular runway to an ethereal score Katrantzou had commissioned from her friend and fellow Greek, the Oscar-winning electro-jazz composer Vangelis. As they passed, I spotted a few materials Mandriota had selected at PV: the organza base on which the stamp motif was printed for swishy day dresses; a fine transparent plastic that Mandriota had pleated in Japan, layered over geometric shifts; a white tulle embroidered with cascading wildflowers for a romantic trapeze midi. The audience cheered as Katrantzou skipped around the runway and took her bow, and the next morning, the critics laid on the praise. Vogue.com: "a walking *wunderkammer* of a collection." The *New York Times*: "opulent mosaics of print and polygons." *Women's Wear Daily*: "fun," especially "fantastic pieces" like the "shimmery floor-length gown with a perfume bottle picked out in sequins down the front" and "flowing nylon dresses . . . printed with famous works of art."

But before any of those reviews were posted, Katrantzou's show guests had uploaded pictures and video clips of the looks on social media, often live. And design teams for fast-fashion brands had perused those images, noted the number of "likes"—an instantaneous, and free, market study—and chosen which designs they would steal, loosely reinterpret, and produce offshore for pennies apiece. (As I walked out of the show, a top online retail executive mused: "I bet Topshop is already working on that butterfly print.") Katrantzou's design work would result in global trends, but she would have no say or stake in the matter.

"It takes three months to produce the forty prints we develop each season," Katrantzou told me. And it takes a click of a smartphone camera to loot them from her. That hurts her business, of course. But it also damages "the whole collective of designers working with digital print, because copying digital print has become so easy," she said. As she well knows, the moment we stop protecting artists and their work—be it words, images, or design—there will be less organic creation, fewer new ideas.

In a matter of weeks, the faux Katrantzous would be churned out in chintzy fabric by poorly paid workers along a fragmented global supply chain and appear in stores retailing for less than one hundred dollars—onetenth, or less, what the far more intricate and luxurious originals would command.

Those tons of garments would be worn briefly, then jettisoned. What remained on the rack for more than a week or two would be marked down, and marked down again—to as little as $3.99—looking sadder and limper with every rejecting whisk of the hanger. Eventually, management would yank the leftovers from circulation and shred or burn them.

This is how the fashion business has functioned on a grand scale for

250 years: creative thievery, indifference for others, corruption, pollution. Ever since an English entrepreneur decided that faster was better.

NOBODY REALLY LIKED Richard Arkwright. A barber and wigmaker by training, he was pompous, litigious, and generally repulsive. "A plain almost gross, bag-cheeked, potbellied Lancashire man," Scottish historian Thomas Carlyle wrote in 1839, "with an air of painful reflection, yet also copious free digestion."

Worse, Arkwright had the habit of stealing other people's ideas—such as Lewis Paul's carding machine and James Hargreaves's spinning jenny—improving them, and using them to his own profit. (Several of his patents were later challenged in court.) In 1771, he pulled together some of these newfangled machines and opened the world's first water-powered textile mill, in Cromford, Derbyshire. With that, Arkwright kicked off the Industrial Revolution—the transition from handmade to machine-made—giving rise to the factory system we still rely on today.

*Clack-clack-clackety-clack* the machines roared, shaking the five-story building, cotton filaments filling the air like a snowy fog. The days were long: thirteen-hour shifts with two short breaks for meals; the mills stopped only for one hour. Workers lived in Arkwright-built brick row houses on the factory grounds and attended an Arkwright-built church. At first, there were two hundred workers; within a decade, a thousand. Local textile factory owner William Radcliffe observed that between 1770 and 1778, "complete change had been effected in the spinning of yarns . . . Wool had disappeared altogether . . . Cotton had become the universal material." By 1790, Arkwright owned nearly two hundred mills throughout the country, and Manchester had become known as "Cottonopolis."

In 1810, a prominent Boston businessman named Francis Cabot Lowell traveled to Europe, ostensibly for a health cure. In truth he was there to nick Arkwright's system. In one of history's greatest feats of industrial espionage, Lowell toured Manchester's mills, memorized the power loom mechanics, returned to Massachusetts, and reconstructed the machines. Three years later, he opened the Boston Manufacturing Company on the Charles River in Waltham, just west of the city, to spin and weave American cotton harvested by slaves.

With the advent of the lockstitch sewing machine in the 1830s, ready-made garment output sped up. But demand remained limited—many people still made their own clothes. Then the Civil War broke out. Both the Union and Confederate armies suddenly needed sturdy uniforms in standard sizes—meaning ready-to-wear—run up fast on those newfangled sewing machines. Factories opened or expanded to meet the demand. The troops so liked the convenience and fit of their regimentals that, after the war, they sought out street clothes made in the same manner. Manufacturers responded by producing menswear, and then women's wear, en masse. It was the genesis of the American apparel industry.

Garment manufacturing in the United States in those early days was divided neatly into two categories: less refined pieces, like work clothes and undergarments that were produced in large, standardized runs at big factories in Massachusetts and Pennsylvania; and stylish, high-quality clothing known as "fashion" that was cut and sewed in smaller quantities in workshops on New York City's Lower East Side.

Why New York? It was America's busiest port, where European wools and silks arrived; the nation's financial center, with bankers eager to invest in the ever-growing garment industry; and the premiere immigration point, with thousands of Europeans disembarking weekly, looking for jobs. A good many were Jews from Hungary, Russia, and what is now Poland—countries where needlework was a respected tradition. In the late nineteenth century, more than half of the Lower East Side's residents were employed in apparel production, and three-quarters of those workers were Jewish.

Most of what they made was inspired by—or copied from—what the Paris couture houses were showing. The most influential was Worth, established on the rue de la Paix in the 1850s by English transplant Charles Frederick Worth—the man generally regarded as the father of modern couture. Until Worth came on the scene, women went to their couturiers and commissioned dresses to their specifications. Worth upended that system by designing "collections" of looks, which he presented to his clients—including the influential tastemaker Empress Eugénie. Then he took orders and produced each gown to measure. His silhouettes appeared in fashion magazines and set trends; he gave us the bustle. The trickle-down system of fashion design, atop which Katrantzou now sits, began with Worth.

As the New York City apparel industry grew, manufacturing spread northward, to appealing new steel-framed loft buildings in midtown Manhattan. The Garment District, as the quarter became known, stretched from Thirtieth Street to Forty-Second Street and from Fifth Avenue to Tenth Avenue, with newly opened Pennsylvania Station smack in the middle, allowing for out-of-town retailers to visit showrooms easily. Business migration was swift: in 1931, the New York Garment District had more apparel factories than anywhere else in the world.

Except for a brief dip early in the Depression, American retailing flourished throughout the 1930s. "Everybody dressed up," New York designer Bill Blass recalled some six decades later. And he wasn't exaggerating: it was a far more formal time, when men and women alike couldn't imagine stepping out of the house without a proper hat. "Some women spent all day in fitting rooms; you put on clothes for lunch, clothes for cocktails, clothes for dinner, whereas today you go to work, lunch, and dinner in the same goddam black pants suit." They shopped at the grand department stores like Macy's and Bergdorf Goodman in New York, Neiman Marcus in Dallas, Selfridges & Co. and Harrods in London, the Galeries Lafayette and Le Bon Marché in Paris, and specialty retailers like Hattie Carnegie at 42 East Forty-Ninth Street.

Miss Carnegie had a discerning eye. In her town house boutique, she sold Paris originals, as well as her own knockoffs of those Paris designs, which retailed for $79.50 to $300 apiece. "Black sheaths caught behind, just below the knees, with pink roses" and "straight, skinny sheaths with full peplums," the *New Yorker* magazine reported in 1941. Joan Crawford would simply wire: "Send me something I'd like." Miss Carnegie also offered an in-house line called Spectator Sports, made in suburban Mount Vernon, with dresses priced at an affordable $16.50 apiece.

Blass landed in Manhattan in 1939, a wide-eyed seventeen-year-old from Fort Wayne, Indiana. "In those days, the excitement of an expensive Broadway opening was rivalled by the free experience of walking up and down Fifth Avenue on Thursday nights, when the department stores unveiled their new windows," he recalled. "You started at Altman's, at Thirty-Fourth Street, and walked to Bergdorf's, at Fifty-Seventh, with a side trip over to Hattie Carnegie, on East Forty-Ninth Street. The window designers vied with one another in originality and outrageousness. Bonwit Teller even hired famous artists like Dalí to decorate the windows."

All toned down during World War II. Gone were peplumed sheaths. "We wore suits," devoted fashion follower Olivia de Havilland once told me. "You married in a suit." Factories turned their attention to uniforms and other wartime necessities. But in the postwar economic boom, American apparel manufacturing returned to fashion with zeal: the Garment District alone counted two hundred thousand women's wear workers, and they produced 66 percent of all US clothing. Blass was one of them—"expected," he said, "to keep [his] head down," and "be grateful that [he] had been given the chance to design $79 copies of Dior dresses."

By the close of the 1950s, Manhattan apparel manufacturing jobs were migrating to the Bronx, Queens, and Brooklyn; upstate, to Rochester; to Pennsylvania; and to Chicago. It was, in effect, a domestic version of offshoring. The reason was economic: with the rise of real estate and labor costs, a dress cost 17 percent more to produce in a New York City than in nearby northeastern Pennsylvania. Manhattan's apparel workers felt the impact: between 1947 and 1956, their earnings dropped 20 percent.

Companies that remained headquartered in the Garment District radically changed the way they conducted business. Fabric was cut in Manhattan work-rooms and trucked to factories outside of the city, where it was sewed into clothes. The finished items were then trucked back to Midtown showrooms and city warehouses and sold to retailers. President Dwight D. Eisenhower's newly opened interstate highway system made the transport easy. But the scheme was ridiculously complicated—all in an effort to pinch pennies. Nevertheless, it stuck, and was the naissance of today's utterly fragmented global supply chain.

As the act of making apparel began to leave the Garment District, it was replaced by something far more creative: the design of *fashion*. Blass and his confreres opened studios on or near Seventh Avenue and turned to neighborhood factories to produce their work, giving the Garment District a boost. Midtown workers wheeled racks of finished clothes down city sidewalks to the showrooms and shipping depots. In 1973, four hundred thousand were employed there—double the 1950s peak. In need of more space, New York's apparel manufacturing business expanded back downtown, specifically to Chinatown, where real estate prices and labor—now immigrants from Hong Kong, with managerial as well as sewing skills—were significantly cheaper. In 1965, there were 35 Chinese-owned workshops on the Lower East Side; in 1980, there were 430, employing some 20,000. In all, 70 percent of the clothing that Americans bought in 1980 was made in the United States.

Then politicians stepped in, and everything changed.


THE NORTH AMERICAN Free Trade Agreement, or NAFTA, was first floated by Ronald Reagan during the kickoff of his 1980 presidential campaign. He saw a "North American accord," as he called it, as a common market "in which the peoples and commerce of its three strong countries flow more freely across their present borders than they do today."

Trade deals—and particularly textile and apparel trade deals—were not new.

After World War II, the US government helped rebuild the Japanese textile industry at the urging of the American cotton-farming lobby; President Harry S. Truman instituted a trade-not-aid policy, with low tariffs. By the late 1950s, the US textile industry began to feel pain from those low-cost imports from Japan, as well as from South Korea, Hong Kong, Singapore, and Taiwan—the Asian "Tigers," as they were known, for their vigorous economies fueled by exports. Washington responded—not only during the Eisenhower years, but for decades to follow—with higher tariffs, and complicated quotas and exemptions.

Even with those higher tariffs, fashion executives realized it was still cheaper to manufacture overseas than at home, and they began to outsource to Asia a bit. Turnaround time was slow—transport by ship took several weeks—but profit margins were substantially greater. In 1960, about 10 percent of the women's wear sold in the US was imported. By the mid-1970s, Hong Kong had become the world's largest apparel exporter, specializing in low-end Western clothing.

"Never!" New York–based women's wear designer Liz Claiborne blasted when her business partner, Jerome Chazen, first suggested offshoring. "How can we possibly control work that is done ten thousand miles away?"

But Chazen insisted that he couldn't find enough manufacturing capacity in the United States to meet the booming demand. Before joining Claiborne, he had worked as a buyer for Winkleman's department store in Detroit and sourced some items in Asia. Maybe that could be a solution, he thought. "My comfort level with the idea was pretty high," he recalled in his memoir years later.

Claiborne's clearly was not. So Chazen suggested they put a small order of a difficult blouse in a Taiwan factory, as a test. "Some weeks later that first lot arrived by air in our offices, and Liz was blown away," Chazen wrote. "It was nicer than anything we had done domestically, and it cost far less than what we had been paying." Good quality, low overhead: Chazen had found the magical formula for financial success.

Before long, Liz Claiborne Inc. was sourcing most of its apparel in Asia— an arrangement that required overhauling the company's production timetable. "Merchandise had to be ordered at least six months before we could expect to ship the garments to our customers," Chazen explained. To achieve that, "Liz and Art would go to Hong Kong every two to three months, and they would hold court in their suite at the Peninsula Hotel," the region's most luxurious lodgings, recalled Liz Claiborne's senior vice president for manufacturing and sourcing Robert Zane. "And they wouldn't leave until the job was done, which meant that the next season's work was designed and production was arranged."

Then came the Reagan Revolution, and with it—as promised on the campaign trail—an economic agenda that was heavy on free trade. Often apparel deals were convoluted—like when the US allowed unlimited quotas for Caribbean countries exporting clothing to the American market, provided that the fabric had been woven and cut in the United States. But they also encouraged brands to join Liz Claiborne offshore.

The resulting job flight so worried US unions that they successfully lobbied Congress to designate December 1986 "Made in America Month." The joint resolution stressed "the importance of buying American" and warned that the glut of imports could permanently reduce the country's production capacity.

Unions and trade groups also launched themselves into an affair that would be their undoing: "quick response" manufacturing. QR, as it was known, was an efficiency system developed by the US Apparel Manufacturing Association in the mid-1980s to compete against imports from low-cost labor markets. At the time, manufacturing experts estimated that the American apparel industry lost $25 billion a year simply through the sort of inefficient business practices honed by Claiborne: because clothes were produced so far away, designers conceived collections a year in advance, and stores placed their orders six to eight months ahead of deliveries. Retailers guessd what might be hot, and if they were wrong, they were stuck with inventory that had to be sold at reduced prices, or scrapped. An industry built on selling new ideas was in fact pushing old ones.

With QR, brands and retailers would test looks with focus groups to see what was successful *before* submitting production orders; said initial orders would be smaller and more frequent; and reorders would be placed only when sales data indicated a need. The intention was to cut inventory levels, amp up inventory turns, and avert leftovers and cut-rate sales. The pipeline would run more efficiently, costs would be lean, and there would be less waste, fewer losses. QR would give customers what they wanted, where they wanted it, when they wanted it.

The management consulting firm Kurt Salmon Associates was brought in to help factories put the plan into action. Investment was steep: a minimum of $100,000 for a small-sized factory. And return was slow: it would take a year to fully implement the system.

In 1987, the Congressional Office of Technology Assessment conducted a study on QR (*The U.S. Textile and Apparel Industry: A Revolution in Progress*) to see how, if fully embraced, it would change manufacturing. The predictions must have sounded like an Isaac Asimov story at the time, but they were eerily prescient:

> Chemicals and robots make our clothing, rather than cotton and
> the sewing machine. Manual labor has been virtually eliminated
> in the production of textiles and apparel, except in design
> and equipment maintenance. Few communities are known as
> "textile towns."
>
> Fiber companies, textile producers, apparel manufacturers,
> and retailers are tied together through sophisticated
> communication networks, and react almost instantaneously to

market trends. Customers enjoy more products tailored to their specific tastes, and find a greater range of styles and sizes in stock. However, demand for blouses and slacks may not be as great as demand for textiles and fabrics used in road construction and rocket ships.

A proliferation of export incentives and import protections among nations of the world has made public policy nearly as important as traditional economic forces. "Made in the U.S.A.," when it appears on a label, may not ensure that all stages of production have occurred within U.S. borders. The domestic industry is comprised of large multinational corporations and small contract shops. Mid-sized firms, the backbone of the industry for two centuries, have all but vanished.

A number of manufacturers decided not to make the investment and embrace QR. So, complete fulfillment of that prophecy would remain decades off. Enough factories did implement the new system, however, to trigger a spike in American production and a drop in imports.

But how long would it last? As a 1990 Harvard Business School report questioned ominously: "Could foreign competitors potentially use Quick Response concepts to once again out-compete domestic producers?"

It was "certainly not unthinkable."

As a matter of fact, in La Coruña, a port town in northwestern Spain, fashion executive Amancio Ortega Gaona was, at that moment, pondering how he could adapt QR to his domestic midrange clothing company, Zara.

Ortega is a lifelong garmento: the son of a railway worker and a housemaid, he entered the rag trade in 1949 as an errand boy for a local shirtmaker. In 1963, he launched Confecciones Goa—his initials, backward. It specialized in wholly unsexy housecoats. In 1975, he and his then-wife Rosalía Mera opened a fashion boutique called Zorba on La Coruña's tony shopping street. When they discovered there was a café Zorba in town, they renamed their place Zara.

For its clothes, Ortega followed the old-school ready-to-wear model—seasonal collections of trendy knockoffs all produced in Spain. It worked well enough—one store grew to eighty-five in Spain by 1989—and he made a good living. But he wanted more.

QR was the key. If Ortega merged its speedy production practices with retailing, he could rev up *everything*: trends, sales, profits. Since his market was domestic, and distances were short, he could get clothes to stores quickly, sell them quickly, restock quickly. Forget seasons; Zara dropped new styles on sales floors constantly. The regular update beckoned customers to come in more often—and they went home with more. Ortega dubbed his new method "instant fashion." And, with it, he changed the apparel business paradigm.

As demand rose, he began to outsource across the Strait of Gibraltar, in Morocco. Labor was abundant and cheaper than in Spain. And the factories were still close—allowing for easy quality control and rapid delivery. Result: greater profit margins.

Ortega's competitors, such as Gap, Urban Outfitters, H&M, and Benetton, took notice. Like Zara, they swiped silhouettes from top-tier fashion houses, reinterpreted them in lesser fabrics, and offered them at bargain prices to the middle-market consumer. All the brands sped up production and sales to such a degree they became collectively known as "fast fashion." They would reshape the planet.

A DECADE LATER, after Ronald Reagan introduced the idea of unifying the United States, Canada, and Mexico as a "common market," American president George H. W. Bush and Canadian prime minister Brian Mulroney took the first step by signing the Canada–United States Free Trade Agreement. Shortly after, they roped in Mexican president Carlos Salinas de

Gortari and rechristened the pact the North American Free Trade Agreement. Negotiations stretched for several years.

NAFTA would eliminate a majority of tariffs—a boon for American business, supporters argued, since the average Mexican tariff on US goods was 10 percent. It would also create a continental emporium of 360 million consumers, with a total annual economic output of $6 trillion. People would buy more, pushing US factories to produce more. "NAFTA means jobs, American jobs, and good-paying American jobs," President Bill Clinton insisted in September 1993, as he kicked off his campaign to win congressional approval.

Not everyone agreed. Texan billionaire businessman Ross Perot, who ran for president as an independent against Clinton and George H. W. Bush in 1992, predicted that NAFTA would cause American industry to decamp—with "a giant sucking sound"—to Mexico for cheaper labor. Perot flung what the *New York Times* editorial page deemed to be "absurd" figures, such as when he claimed that 85 million Americans could lose their jobs. It was hyperbole, of course. But the core of his argument was spot-on: companies did move offshore—to Mexico and elsewhere. By 2006, NAFTA was responsible for the loss of at least a million jobs—some analysts estimate much more—and had gutted scores of once-virile domestic industries, most notably textiles and apparel.

Even so, governments continued to negotiate trade deals that encouraged offshoring, with no change in oversight regulations or enforcement. In 2001, China joined the World Trade Organization (WTO), the Geneva-based intergovernmental association that regulates global trade. In 2003, the World Bank trumpeted that the elimination of trade subsidies, barriers, and tariffs would raise the wages of 320 million workers above the $2-a-day poverty level by 2015. Three years later, the bank revised that figure: with the offshoring rush to every cheap labor market, only 6 to 12 million would get the salary bump.

Meanwhile, from 2003 to 2013, China's apparel exports to the United States multiplied five-fold, even though the tariff on clothing was as high as 13.2 percent—nearly ten times what most imported items had to pay. This proved how little fashion cost to produce: even with such high tariffs, everyone along the supply chain, including the brand owners, could still make a decent—or more than decent—profit.

What happened?

The global explosion of fast fashion.

Throughout the 1990s, fast fashion, like the rest of the apparel industry, trotted along at a pretty good clip. By 2000, retail spending on clothing and accessories hit approximately $828 billion, or €900 billion, worldwide, evenly split among major markets: the US, 29 percent; western Europe, 34 percent; and Asia, 23 percent. Fast fashion accounted for a major chunk of those sales. In 2001, Zara had 507 stores worldwide and could get clothes from the drawing board to the sales floor in five to six weeks; traditional brands would take six months. That May, Zara's parent, Inditex, listed a 26 percent stake in the corporation on the Madrid stock exchange; Ortega retained more than 60 percent.

The IPO windfall would help underwrite expansion. Between 2001 and 2018—as trade barriers tumbled and globalization hit its stride—Zara rolled out nearly 1,700 new stores, for a total of 2,200, in ninety-six countries. Usually, the boutiques were in proximity to luxury stars such as Louis Vuitton and Gucci, to coast on those glam houses' appeal as well as pull in their deep-pocketed shoppers.

All the while, Inditex remained relatively hidden from view. No interviews. No advertising. No photographs—not even of Ortega; there are very few public images of him. Little is known of his private life: he and his wife Rosalía, with whom he founded Inditex, divorced in 1986; she died in 2013. He remarried in 2001. One of my friends, a former business writer for *Women's Wear Daily*, told me he called Inditex headquarters almost every day for ten years to request an interview; every day he was told, "Not now." But in 2015, Imran Amed, founder of Business of Fashion, a corporation-friendly apparel news web platform, and one of his reporters accepted a trip to the La Coruña headquarters, underwritten by Inditex, and published a flattering piece that finally revealed, as much as the powers in charge would allow, the workings of the company. Amed was astonished by what he saw. "The scale was extraordinary, as was how they thought out everything from the size of the hangers to how you decide to store things," he told me. "They optimize every step."

On the La Coruña campus, a data center hums 24/7 to process information on the company's supply chain, sales, social media, environmental emissions, energy consumption, and more—all to reduce waste of time, resources, and money. "Here you can control the whole world," Inditex's chief communications officer, Jesús Echevarría, crowed to BoF.

Twice a week, Inditex shipped new items to the group's 6,500 boutiques as well as its e-tailing depots. To understand how this nonstop cycle of refreshing and replenishing the stock impacts sales, consider this: shoppers drop into most fast-fashion brands' boutiques four times a year; for Zara, they go seventeen.

If a style doesn't sell out within a week, it's pulled from the sales floor, and manufacturing orders are canceled. If it does seduce shoppers—a phenomenon the data analysts clock instantaneously in La Coruña—the order is re-upped and manufactured in small batches at factories in Spain, Portugal, and Morocco—all a short hop to the distribution center. After a month or so, it will have run its course and be replaced by another hot new look.

Zara's nimble practices make it roughly four times more profitable than its peers and explain why Ortega regularly appears in the top ten of *Forbes*'s annual world's richest list; on occasion, he is number one. As Inditex's largest shareholder, he personally netted $45 billion between 2009 and 2014. He announced his retirement in 2011 and ceded full control in 2017; in 2018, he was still reaping $400 million a year in dividends, and *Forbes* declared he possessed $70 billion, making him the sixth-wealthiest person after Jeff Bezos, Bill Gates, Warren Buffett, Bernard Arnault, and Mark Zuckerberg. In 2017, Zara did just shy of $19 billion in sales.

As before, its competitors followed form, opening hundreds of outlets worldwide. The application of new digital and communication technologies further streamlined operations, tightened manufacturing cycles, and increased output. Between 2000 and 2014, the number of garments produced doubled to 100 billion annually—or, as McKinsey researchers noted, fourteen new garments per person per year for every person on the planet. The supply chain fractured even more. Fabric would be woven and dyed in one locale, cut in another, sewed in a third, with zippers and buttons attached in a fourth. And finishing touches—denim distressing, embroidery—were executed in yet another land. Nearly every step was—and is—contracted or subcontracted; few fashion companies own their factories.

Preferred transport is by sea. It takes time but costs substantially less than by air. To speed up the process and shrink budgets even further, it is rumored that one major fast-fashion company has refitted cargo ships with sewing machines and other production equipment so the clothes can be made *during* the voyage in international waters.

Because fast-fashion companies chase the lowest bids every step of the way, they can slash retail prices without diminishing profits. Between 2000 and 2014, US consumer prices in real terms shot up 50 percent, but clothing prices actually *dropped*. Stores were suddenly teeming with $5 T-shirts and $20 dresses—nearly the same price as Hattie Carnegie's Spectator Sports dresses during the Depression—encouraging shoppers to buy more, more, more. And they did. As McKinsey reported, "the number of garments purchased each year by the average consumer increased by 60 percent." People burned through clothes at an unprecedented rate. "Throwaway clothes" became normal.

Some argue that Zara et al. have democratized fashion by bringing high design to the mass market. Anna Wintour, the editor of American *Vogue*, once told me, "The more people who can have fashion, the better." But it also preys on our insecurities and our increasingly short attention spans. We are prone to a barrage of fashion images—on social media, on television, on billboards, in the press—begging us, taunting us to indulge in what one executive described as a "temporary treasure."

Fast fashion's target audience is young—eighteen to twenty-four years old, a demographic that doesn't hold on to clothing purchases for very long. According to research by the retail consultancy Kurt Salmon, one-third of these shoppers buy a fashion item every fortnight, and 13 percent indulge once a week. They order online, and of the younger slice—the eighteen- to twenty-year-olds—20 percent want same-day delivery. They don the item immediately and pose for selfies, which they post on Instagram or Snapchat. Then they toss it, donate it, or resell it and go shopping some more.

This maelstrom has sucked in every segment of fashion. "Even designer collections are forced to adopt an industrial pace and scale . . . [to] compete against fast-fashion behemoths like Zara and H&M," French designer Jean Paul Gaultier lamented back in 2016. The hamster-wheel cycle drove him to abandon ready-to-wear in 2015, after a forty-year run, and focus solely on made-to-order couture. "The system doesn't work . . . There aren't enough people to buy them. We're making clothes that aren't destined to be worn," he said. "Too many clothes kills clothes."

NEVERTHELESS, we keep buying them, and discarding them, which incites fast fashion to carry on stealing other designers' ideas. In the early 2010s, Mary Katrantzou was so fed up with the Irish retailer Primark's thieving, she got a lawyer. She wasn't the first to go after a fast-fashion brand for copyright infringement; in 2011, *Forbes* published a report stating that Forever 21 had been sued approximately fifty-one times. All the cases were settled, usually for undisclosed sums. On occasion, however, financial awards have been revealed, and they are a pittance compared to the lucre fast-fashion brands amass. In 2011, Primark paid $140,000 to textile designer Ashley Wilde. That same year, it raked in more than £309 million, or roughly $480 million, in profit.

Even if the payoff was small, Katrantzou thought it worth pursuing—her primary objective was to thwart Primark's dissemination of the knockoffs. "After months and months of discussion—they were ignoring us—the response we got was that it wasn't actually referencing my work. It was referencing a Brazilian designer who I'd never heard about," she told me. "I went and looked at the Brazilian designer: he had copied me. A flat-out copy!" It seems that, to add a layer of legal protection, fast fashioners copy lesser-known copiers.

"This all lasted eight months," she continued, "and in the end, Primark agreed to remove all merchandise. But by that point there was no merchandise." She looked at me saucer-eyed and sighed. "They had sold it all."

# TWO

# The Price of Furious Fashion

ON THE TOP FLOOR of the Bendix Building, a run-down eleven-story Gothic Revival office tower in the heart of Los Angeles's Fashion District, through metal gates left slightly ajar, I spied workers hunched over machines in badly lit rooms, sewing clothes. Mounds of fabric cluttered the linoleum floor; threads, scraps, and dust bunnies were everywhere.

Suddenly, doors began to slam shut, one after the other. *Boom! Boom! Boom!*

"Wow, that was quick," observed Mariela "Mar" Martinez, an organizer for the L.A.-based nonprofit Garment Worker Center. Someone had recognized her and alerted the neighbors.

We descended to the eighth floor; its workshop doors were already closed and locked. The tip-off went building-wide. At the end hallway, we looked across the street to the Allied Crafts Building, another of the dissolute early-twentieth-century downtown towers that houses sweatshops. Its art deco façade was crumbling. Several windows were whitewashed opaque. A few of the rotten sashes were open a crack—enough for us to hear the sewing machines clattering.

We took the Bendix elevator back to the street level and walked over. In the lobby was a check-cashing booth and a Latino man using a pay phone. Martinez explained that most L.A. sweatshop workers are Latino, and most sweatshop owners are Korean. We climbed the stairs to the third story. Windowpanes were broken. A thirtyish man—a manager, perhaps—sat on a step of the rusty fire escape, smoking. "Would you go down this?" Martinez asked me. With skinny cables and hundred-year-old wall fasteners, it appeared it would buckle under the weight of more than a couple of people. And once you got down to the second floor—where it stopped—you'd have to jump into a Dumpster.

"I call this the white noise of Los Angeles," Martinez said as we made our way back to the street. "No one sees it, or acknowledges it, but it's here."

Today, Los Angeles is America's largest apparel manufacturing center. The sector began in the early twentieth century, when local knitting mills started to specialize in swimwear—brands at the time included Cole of California and Catalina. It continued to grow after World War II as the "California Look"—a casual-chic silhouette in lighter fabrics—became popular across the country. Finally, in the early 1990s, Los Angeles supplanted New York as the US fashion production capital—rising Midtown real estate prices and NAFTA were a one-two gut punch to the Garment District. In 2017, Ilse Metchek, president of the California Fashion Association, told me the local industry's annual revenue was about $42 billion. Martinez estimated there were 45,000 workers producing apparel in Los Angeles. About half were on the books and paid at least the California minimum wage, which was then $10.50 an hour.

The other half were undocumented, sewing garments for US-based brands in clandestine factories for as little as $4 an hour. No overtime. No health benefits. Horrendous conditions. Yet the large, middle-market brands who source from these sweatshops herald that their apparel is "Made in the USA," as if such a statement automatically confers authenticity and integrity, as well as a superior quality, to clothes produced offshore. It's a corporate marketing tactic to pander to consumers' patriotism while flagrantly breaking US labor laws.

Domestic sweatshops have always existed. In Richard Arkwright's day, nearly every factory was a sweatshop. The same was true on New York's Lower East Side in the late nineteenth and early twentieth centuries. When unions and labor laws banished them, they went underground. Run by organized crime, domestic sweatshops became hidden hubs for human trafficking and money laundering. On occasion, the discovery of one makes the news—and usually the scene is horrifying. In 1995, state and federal agents raided a clandestine apparel factory in the L.A. suburb of El Monte that was surrounded by barbed wire and spiked fences and guarded by sentries. Inside, they discovered seventy-two enslaved Thai workers, $750,000 in banknotes and gold bullion, and records showing bank transfers of hundreds of thousands of dollars in cash.

With the current backlash against globalism and the accompanying protectionist calls to buy American, domestic sweatshops have grown more pervasive, especially in L.A., because of its large undocumented immigrant population. According to a UCLA Labor Center study released in 2016 and coauthored by Martinez, 72 percent of Los Angeles garment workers stated that factories were dirty; 60 percent said they were poorly ventilated and led to respiratory ailments; 47 percent reported disgusting restrooms; and 42 percent said they had seen rats. According to the investigation, the brands allegedly producing in such conditions included Forever 21, Wet Seal, Papaya, and Charlotte Russe.

In 2016, the US Department of Labor charged that these and other Southern California apparel manufacturers violated basic federal protections, such as paying minimum wage and overtime, 85 percent of the time and ordered the suppliers to pay $1.3 million in back wages and damages. (Forever 21 and Russe later stated they took labor issues "very seriously"; Forever 21 added,

"These entities are completely independent of Forever 21 and make independent business decisions.") Most of these suppliers were located "right in the heart of the Fashion District—twenty blocks from City Hall," UCLA study coauthor Janna Shadduck-Hernández later said.

That's what drew Martinez to the fight. A young woman in her twenties, she grew up a few miles from the Garment Worker Center, in South Central. Her parents were employed in the city's legitimate garment industry—her father running embroidery machines, her mother cutting samples for fashion brands. She became a human rights activist while in high school and, as an undergrad at Brown University, joined the United Students Against Sweatshops (USAS), a youth organization committed to change via campaigns and boycotts. After Brown, she returned to Los Angeles and joined the Garment Worker Center as its organizing coordinator. Two afternoons a week, she meets with workers in the center's windowless offices in a beat-up lowrise on Los Angeles Street and listens to their grievances.

Most common is "wage theft": when bosses pay workers significantly less than the state or federal minimum wage. Usually, she'll contact the employer directly and try to negotiate a deal. If the case is particularly egregious, she'll reach out to state and federal agencies, such as the US Department of Labor's Wage and Hour Division, which will launch an investigation that may result in a raid. Martinez accompanies agents on the sweeps, and, sometimes, she says, she spots labels of brands who market their clothes as "sweatshop-free." When caught, the brands often claim they had no idea their "approved" contractors were subcontracting to sweatshops. Subcontracting is endemic in the apparel industry, creating a fractured supply chain in which workers are easily in jeopardy.

Martinez or government officials file a claim for lost earnings—the difference between minimum wage and what workers are actually paid. Everyone along the supply chain—subcontractors, contractors, brands, retailers—bucks responsibility. Factories "will close shop, or reopen under a different name," she said. "They use fake IDs, or the employer might not be the person who's on the registration. That will break a case."

When Martinez does manage to recuperate monies, it's usually "not even half of what's actually owed," she said. "We have wage claims that are fifty thousand dollars, not counting the penalties, and we'll recover five thousand dollars, ten thousand dollars max. And that's when the worker has a representative. Without a representative, the contractor will offer one hundred dollars, two hundred dollars, and a lot of people take it because it's from nothing to something."

After such meetings, Martinez continued, "brands tell the contractors, 'Either you fix this or we're taking *all* our work away from you.' They wash their hands of the problem. And if the brand leaves, the contractor has no money to pay the wages owed. I'm not super sympathetic to the contractors, but they are pawns in this system. If every garment worker in L.A. were to file wage claims, that'd be millions owed—millions that went to the pockets of CEOs." And shareholders.

She looked at me sullenly. "Listen, we all know our shit's made in sweatshops," she said. "But we put it in the back of our minds. Nobody cares."


FOR HIS MILLS TO SUCCEED, Richard Arkwright needed cotton. Lots of it. And with the British Empire's global network of trade routes, lots of it was available. In the eighteenth century, the economic cornerstone of Britain's colonies in America and the Caribbean was cotton. Slaves grew it, harvested it, and loaded it onto ships headed for Arkwright's factories in England. As *Communist Manifesto* coauthor Karl Marx later observed, "Without slavery, there would be no cotton. Without cotton, there would be no modern industry."

To spin the cotton, Arkwright needed armies of workers. Hundreds of poor streamed in from town and country. Most were women; men remained on the farm to tend the fields. Before Arkwright, women ran homes and raised children; he turned them into wage earners. He hired the unattended children, too, paying them a fraction of what the adults got.

Friedrich Engels, the son of a German textile tycoon and apprentice in the Manchester cotton industry, was so appalled by what he witnessed there, he penned *The Condition of the Working Class in England*, published in 1845. Mill workers were "robbed of all humanity" and trapped in unimaginable poverty, he wrote. The average life expectancy of Manchester's working class was seventeen. Epidemics—cholera, smallpox, scarlet fever—were "three times more fatal" in Liverpool than in the countryside, and alcoholism surged, with "people staggering . . . [and] lying in the gutter." Half of Britain's factory hands were women; they cost less, and were more docile, than men. And by the 1840s, just as many workers were under eighteen years old as over—equally divided between boys and girls. Starting age was typically eight or nine.

The endless hours Cottonopolis factory children spent on their feet stunted their growth and caused ailments such as chronic back pain, varicose veins, and large, infected ulcers on their legs. In the summer of 1843 alone, the *Manchester Guardian* reported a boy died of lockjaw after his hand was pulverized between wheels; a child was caught up in a cogwheel and crushed to death; and a girl was snared by a strap and spun around the machinery fifty times. Not surprisingly, child workers tried to escape. Some hid in storage rooms to sleep—only to be found by the bosses and beaten. The tales in former factory boy Robert Blincoe's 1832 memoir were so harrowing, the book is believed to have inspired *Oliver Twist*.

The adults were equally imperiled. Standing for hours in contorted positions, manipulating heavy machinery, and hauling burdensome loads deformed women's pelvises, which caused them to miscarry or die in childbirth. Rape on the job was common, with pregnancy rates spiking when factories ran at night. The fibrous dust workers inhaled on the plant floor induced respiratory infections, asthma, "blood-spitting," and consumption. Some were felled by injuries such as "the loss of the whole finger, half or a whole hand, an arm, etc., in the machinery." Mill work, Engels declared, was a new mode of enslavement.

Public outcry eventually pressured the British government to pass a series of laws regulating factory conditions and wages. But the new regulations were ignored since, as Engels pointed out, fines were "trifling" compared to "the certain profits." Most galling, however, was what he viewed as the hypocrisy of England's wealthy. With their philanthropic endeavors, they hailed themselves "as mighty benefactors of humanity," he wrote, when, in fact, such charitable donations simply returned to "plundered victims the hundredth part of what belongs to them!" He believed that was the worst crime of all.

WHEN THE GARMENT INDUSTRY moved to America in the nineteenth century, the labor abuses followed. There, too, much of the philanthropy was ineffective window dressing. But there were exceptions. In 1890, two young, wealthy progressives—Josephine Shaw Lowell, a Civil War widow whose late husband was the nephew of Francis Cabot Lowell, and Maud Nathan, a Sephardic Jew married to a stockbroker—formed the Consumers' League of New York City, a nonprofit advocacy group of middle-class women dedicated to improving employment conditions in the local apparel industry. Their motive was both civic-minded and personal: they were troubled by the reports of exploitation of female and child workers, but they also worried about contagions embedding in their clothing.

The US House of Representatives subsequently launched an inquiry into America's apparel industry and heard overwhelming evidence for the need for reform. But nothing changed. So activist Florence Kelley made the abolition of American sweatshops her crusade. As the first general secretary of the National Consumers League—a nationwide nonprofit founded in 1899 that united local consumer organizations—she argued that modern mechanization and streamlined distribution, rather than anemic wages, were the most effective ways to reduce production costs. Indeed, she believed the sweatshop system *increased* costs because it discouraged owners from updating their machinery. She called for boycotts, stating: "If the people would notify [Chicago department store retailer] Marshall Field . . . and others that they would buy from them no clothing made in sweatshops, the evil would be stopped."

In 1899, the National Consumers League introduced the "white label"— a garment tag that certified the manufacturer had respected state employment and safety laws as well as the League's standards. The white label empowered shoppers, pushing them to think ethically about their purchases. "We can have cheap underwear righteously made and clean; or we can have cheap underwear degradingly made and unclean," Kelley declared. "Henceforth we are responsible for our choice."

Some retailers balked, but not Philadelphia department store magnate John Wanamaker. He joined the League's campaign to improve factory conditions, promoted white-label-approved clothes in his store, and had window dressers fill the Broad Street vitrines with displays that demonstrated the differences between sweatshops and white-label-approved factories. Photographs of the window exhibit later toured international trade fairs. Within five years, sixty American manufacturers had qualified to use the white label in their clothing.

Still, many factories were constructed on the cheap and often violated health and safety codes. A common infraction was locking emergency exits to prevent employee theft. Such conditions courted disaster—like the Triangle Shirtwaist Factory fire in 1911. Fleeing workers ran onto the rickety fire escape, and it collapsed. Dozens leapt from windows and the rooftop— many with hair and clothing ablaze. In all, 146 employees died—123 women and 23 men. It was New York City's worst workplace disaster until September 11, 2001.

To fight back, in stepped Frances Perkins. A dynamic advocate for worker rights, in 1910, she became the executive secretary of the New York City

Consumers League, where she worked side by side with Florence Kelley. After Triangle, Perkins joined New York State's Industrial Commission, which regulated factories. In the 1930s, President Franklin D. Roosevelt named her secretary of labor—the country's first female cabinet member. During her twelve-year tenure—the longest in that post—a multitude of landmark acts were passed and agencies created, including the Public Works Administration; the Social Security Act, which established unemployment, welfare, and retirement benefits; and the Fair Labor Standards Act (FLSA), which set forth the country's first minimum wage, guaranteed overtime payment, banned child labor, and instituted the forty-hour workweek. With the FLSA, American manufacturing cleaned up and moved into its golden age.

Except the Garment District. It remained "a place of Dreiserian amounts of soot and lint," Bill Blass recalled. "Manufacturers did their best to foster an atmosphere of contempt . . . Even after those of us came home from the war and were still in uniform, we weren't permitted to ride in the same elevator with our employers. We were backroom boys in the grubby business to end all grubby businesses—Seventh Avenue."

AND SO IT REMAINED, until apparel manufacturing moved offshore post-NAFTA and most of those workshops folded. Offshore, the old-style sweatshop system came roaring back to life. In developing economies, labor laws were far less restrictive, and there was little or no oversight. Thus why, within six months of Congress passing NAFTA in 1993, the House Subcommittee on Labor Management found itself holding hearings on worker abuses in a Honduran factory where the American women's wear brand Leslie Fay was sourcing clothes.

Leslie Fay had long been a standard-bearer in American fashion. Founded in 1947 by Fred Pomerantz, a cigar-chomping apparel executive who had worked in Manhattan's Garment District from age eleven, Leslie Fay—named for his only daughter—was known for its flattering dresses, sewed by union workers at the Wilkes-Barre plant. By the time Fred retired in 1982, the brand was publicly traded and sold in thirteen thousand department and specialty stores nationwide, with an impressive turnover of $500 million a year.

Fred's Wharton-educated, middle-aged son, John—on the Leslie Fay staff for decades—took the company private through a leveraged buyout—a popular financial business ploy in the boom-boom '80s. Two years later, a second leveraged buyout by former company managers and independent investors netted John Pomerantz a whopping $41 million. With this immense wealth, he and his wife, Laura, became bright lights of the New Society, as the decade's American *über*rich were known.

In 1986, Leslie Fay was listed on the stock market again, and sales reached an impressive $859 million in 1990. John was named chief executive officer; Laura, a former investment banker raised in a retailing family, served as a senior vice president. In January 1993, while on a business trip in Toronto, John Pomerantz received a call from his chief financial officer, Paul F. Polishan. "We got a problem," Polishan said ominously. "Maybe a little more than just a problem."

Small, privately held apparel companies are known to fudge accounting from season to season—counting sales orders, and profits, before they are actually completed and earned—to make end-of-year numbers look better. But Leslie Fay was a publicly traded company, and the numbers weren't simply fudged: the brand reported earning $24 million, when it had actually lost $13.7 million.

When the news broke, the company's stock crashed, shareholders filed a class-action lawsuit, and two months later, Leslie Fay went into Chapter 11 bankruptcy protection. Pomerantz swore that he never knew about the accounting fraud; it was, he said, the initiative of rogue employees. (Polishan was later indicted and eventually did time in prison.)

Executives at the Wilkes-Barre headquarters initiated a cost-cutting plan. Until that time, Pomerantz had resisted outsourcing, because he believed domestic manufacturing, with its faster turnaround time from plant to retail floor, was smart business. He also said he felt a sense of moral duty to keep production at home.

With the bankruptcy, all that good sense and moralizing went to the wayside. Leslie Fay moved production to Honduras—a long way from management's reach in northeastern Pennsylvania. Soon it became apparent that, like so many other American apparel companies manufacturing outside the United States, Leslie Fay's execs had no idea how the brand's clothes were getting made.

They learned, to their embarrassment, from witnesses during the congressional hearings in Wilkes-Barre in 1994. Dorka Nohemi Diaz Lopez, a twenty-year-old Honduran who sewed Leslie Fay dresses and blouses, was brought before them by the National Labor Committee (NLC), a Pittsburgh-based whistle-blowing nonprofit dedicated to halting human and labor rights abuses. Lopez told subcommittee members that girls as young as thirteen were on the plant floor, earning 40 to 50 cents an hour; Leslie Fay's US-based workers were paid $7.80 an hour. Conditions in the Honduras factory were Mancunian. The girls logged twelve-hour shifts or longer. Temperatures often exceeded 100 degrees, and there wasn't any clean drinking water. "The doors are locked," she testified, "and you can't get out until they let you."

Back in Wilkes-Barre, backlash was fierce: children of laid-off workers wrote to Pomerantz, asking him why he took their parents' jobs away; preachers scolded the company in Sunday sermons; displaced workers staged protests; the regional papers ran brutal op-eds excoriating Pomerantz for the move offshore. "We always felt it was like a family," fifty-six-year-old Jeannie Kowalewski, a Leslie Fay machine operator for thirty-eight years, told the subcommittee.

Pomerantz was nonplussed. In a letter to the congressional subcommittee, he wrote, "This is a bogus issue. [Offshoring] low-skilled jobs . . . is what the NAFTA debate was all about, and that debate is over."

MANY AMERICAN APPAREL companies, including household names like Kathie Lee Gifford, J. Crew, Eddie Bauer, and Levi Strauss, faced similar charges. In response, some started drafting "codes of conduct": a list of standards that a company expects its suppliers to respect. *Expects*. None of it is mandatory. All is voluntary. The need for codes of conduct shone a light on fashion's greatest, and seemingly unsolvable, conflict: how to produce for the lowest price possible while ensuring safe, humane conditions and decent pay.

Levi Strauss's executive management committee approved fashion's first code of conduct in March 1992—before NAFTA. Drawn up by a Levi's internal task force called the Sourcing Guidelines Working Group (SGWG), the code was "to ensure the people making our products in contract factories were being treated with dignity and respect, and working under safe and healthy working conditions," the company stated. The SGWG used the UN's Universal Declaration of Human Rights and the International Labour Organization's rules as its guide: no child or forced labor; no gender, race, or ethnic discrimination; legal hours; fair wages; benefits; freedom of association and collective bargaining; and adequate health, safety, and environmental standards for all manufacturers that produced its clothes.

The brand's motivation was questionable. Levi Strauss introduced its code shortly after it canceled contracts with a Hong Kong–owned factory in the American territory of Saipan that had reportedly violated worker rights. The scene in the plant was as bad as in developing nations: seedy dorms, endless hours of overtime, squalid toilets, locked fire exits, all on a compound fenced in by razor wire and patrolled by armed guards. As the facility officially sat on US soil, the companies who produced there, including Levi's, Gap, Ralph Lauren, and Liz Claiborne, were allowed to use "Made in the USA" tags. Days after Levi's put its code into effect, the US Labor Department discovered federal health and safety infractions in more than a dozen

Saipan-based factories owned by the same Hong Kong family and filed suit. Eventually, the factory's owner forked over $9 million in back wages to workers.

To enforce the codes, brands hired independent monitors to conduct audits. As is still the case today, monitors announced their visits in advance, so factories would be cleaned and workers coached on what to say. In some countries, up to half the factories reportedly buffed up their employee records to pass inspections. Monitors had no oversight. Bribery was common. And scandals kept surfacing.

In 2003, American rap stars Sean "P-Diddy" Combs and Jay-Z were caught up in one: clothes for their hip-hop fashion brands Sean John and Rocawear, respectively, were found to be made in Honduran sweatshops. At a Senate Democratic Policy Committee hearing in November of that year, Lydda Eli Gonzalez, a nineteen-year-old Honduran garment worker, recounted, through a translator, the horrors she had experienced at Southeast Textiles (SETISA). Roughly 80 percent of SETISA's output was Sean John; the remaining 20 percent was Rocawear.

The industrial zone where SETISA was located was surrounded by a towering wall, its entrance a locked metal gate guarded by armed sentries. Official hours were 7 a.m. to 4:45 p.m.—at 75 to 98 cents an hour—but there was mandatory unpaid overtime. The Sean John shirts retailed for $40 apiece at American department stores such as Bloomingdale's. The factory produced more than a thousand each day. "Just one shirt would pay more than my wage for a week," Gonzalez testified.

Supervisors would "stand over us shouting and cursing at us to go faster [and calling] us filthy names, like *maldito* [damned] donkey, bitch, and worse," she continued. The temperature rose so high, workers were "sweating all day." Fabric fibers and dust turned their hair "white or red or whatever the color of the shirts we are working on." The drinking water reportedly contained fecal matter. Workers were forbidden to speak. They could only use the restroom once in the morning and once in the afternoon, and before entering, they were searched; normally, there was no toilet paper or soap. Women were subjected to pregnancy tests; if one came back positive, she'd be sacked. All were frisked upon entering the factory each day, and anything found, including candy or lipstick, was confiscated. They were patted down again when they punched out at night.

Combs understood this news could slay his brand and promptly stepped in. Within ten weeks, the National Labor Committee (NLC; later known as the Institute for Global Labour and Human Rights) announced that the SE-TISA factory's production chief and deputy had been let go; overtime was now voluntary and paid; locks had been taken off bathrooms and the guards banished; air-conditioning and a water purification system had been installed. Workers would all be registered for national health care and were permitted to establish a union. There was talk of abolishing the pregnancy tests, too.

Nevertheless, pay remained hopelessly low.

"Really, you work just to eat. It's impossible to save. You can't buy anything. It's just to survive," Gonzalez told the Senate subcommittee. "I'm no better off than I was two or three years ago. We are in a trap."

IN NO PLACE is the trap more inescapable or harrowing as in the People's Republic of Bangladesh.

A sliver between India and Myanmar in the region of Bengal, in 2019, it counted 168 million citizens, approximately one-fourth living below the poverty line. It is the ninth most populous country in the world, the tenth most densely settled.

In fiscal year 2018, 40 million workers produced more than $30 billion worth of "ready-made garments," or RMGs, for export, ranking Bangladesh the number two apparel producer, after China/Hong Kong, according to the WTO. "Eighty-three percent of our foreign currency is coming from this sector," Siddiqur Rahman, president of the Bangladesh Garment Manufacturers and Exporters Association, told me. "Fifty million people depend on the garment industry. Our *economy* is dependent on it." The government planned to double output in five years.

Bangladesh's garment industry is relatively young—born in the 1970s, following the country's war of independence from Pakistan. South Korea had maxed out its American apparel and textile quotas at the time, so manufacturing entrepreneurs built and outfitted factories across Bangladesh's rural landscape. As with every new garment industry locale before it, impoverished young women flocked—or were dispatched by their families—to these places for jobs. With rock-bottom wages and unconscionably long hours, Bangladesh became the cheapest place to produce apparel—a new Manchester.

Suppliers erected shoddy plants by the thousands, often without permits or such basic safety precautions as grounded wiring or fire exits, but always with top-notch security—to keep the workers in and the pilfering down. Bangladeshi factories were far removed from the contracting brands' headquarters, and these workhouses went unseen and unknown.

Change would be driven by NGO advocates like Judy Gearhart, the head of the International Labor Rights Forum (ILRF), a nonprofit human rights organization in downtown Washington, DC, founded in 1986 to advance "the dignity and justice for workers in the global economy." No-nonsense yet affable and unquestionably dedicated to the cause, Gearhart began defending workers in 1992, in Mexico, during the NAFTA fight; she joined the ILRF as executive director in 2011.

The ILRF has long had a presence in Bangladesh to combat child labor. But on April 11, 2005, Spectrum Sweater Industries Ltd., a poorly constructed nine-story factory in the Dhaka suburb of Savar, imploded shortly after midnight, killing sixty-four and wounding eighty. After that, Gearhart told me, "our apparel industry work really deepened. We started tracking factory fires and collapses and working more closely with the Clean Clothes Campaign and Worker Rights Consortium to campaign against companies caught with product in any of those factories."

The ILRF's strategy is three-pronged: advance legal and policy reforms; demand greater corporate accountability; and support and strengthen the influence of workers and local worker organizations. But resistance remains. The garment industry generates an enormous cash flow for the Bangladeshi government, and not only on a revenue level. In 2018, 10 percent of Bangladeshi parliament members owned garment factories and 30 percent had family members who were owners, the ILRF's director of organizing and communications, Liana Foxvog, told me. "So, you can imagine the collusion," she said. And the corruption. And the graft. Which, as in New York a hundred years ago, leads to disaster.

IN DECEMBER 2010, the ten-story That's It Sportswear garment factory outside Dhaka caught fire—despite having just passed inspection by representatives of Gap. The scene was familiar: locked exits, workers defenestrating themselves. More than one hundred were injured, and twenty-nine perished. They were not alone; between 2006 and 2012, more than five hundred Bangladeshi garment workers died in factory fires. Since That's It Sportswear produced clothes for such big names as Gap, Tommy Hilfiger, and Kohl's, the story made international papers, and there were calls for reforms.

Worker unions and NGOs sat down with brands to discuss factory safety and hammered out a legally binding accord called the Bangladesh Fire and Building Safety Agreement. It sat unsigned until the winter of 2012, when ABC News in New York picked up the That's It Sportswear story, more than a year after the inferno, and questioned the designer Tommy Hilfiger and his chief executive on producing in such a firetrap. Only then did PVH Corp., which owns the Tommy Hilfiger, Calvin Klein, Van Heusen, IZOD, and

Arrow brands and holds licenses for Michael Kors, Sean John, and Speedo, agree to sign the convention. Six months later, the German retail chain Tchibo also joined. But no one else did, and the agreement would not go into effect until four companies were on board.

On a November evening, eight weeks after the safety agreement stalled, twenty-three-year-old Sumi Abedin was sitting at her sewing machine on the fourth floor of the nine-story Tazreen Fashion factory in the Dhaka suburb of Ashulia when, she recalled, "a man came up and shouted that there was a fire."

Her manager and supervisor assured everyone that nothing was amiss.

"There is no fire, just go back and keep working," they told the workers and locked the door.

Fire alarms sounded. Supervisors and security guards insisted it was only a drill and told them to stay on task.

"After five to seven minutes, I smelled smoke," Abedin remembered. "I ran to the doors, the stairs, and found that [they were] padlocked . . . Smoke was coming from the downstairs." She managed to get to the second floor, but that was it," she said. The staircase was "blocked by the fire."

More than 1,100 were trapped inside. Doorways and stairwells that were open were narrow, and fire escapes were few and dilapidated. Workers tried to remove security bars from windows. One succeeded. He jumped. Then another jumped.

"Then I jumped," Abedin said.

She broke her arm and foot. The coworker who jumped with her died on impact.

In all, more than 200 were injured, and at least 117 died, nearly half burned beyond recognition. It was the worst apparel industry accident since the Triangle Shirtwaist Factory fire a century earlier. Investigators later found labels, clothing, and paperwork proving Sears, Walmart, and Disney had all produced there. All three claimed Tazreen was an unauthorized supplier.

Remarkably, even after Tazreen, which got enormous play in the global media, "brands still didn't feel compelled" to sign the Bangladesh Fire and Building Safety Agreement, ILRF's Liana Foxvog told me. That made shady garment factory owners like Sohel Rana feel positively omnipotent.

SOHEL RANA WAS A THUG. In his midthirties, the Bangladeshi operative was known for his bullying business tactics as much as his swagger. He'd tool around Savar on his motorcycle, his biker gang in tow. A political operator, he had public officials and the police in his pocket, allowing him to deal drugs and rough up perceived enemies with impunity. He made his money with his father, who sold the family's country land in the late 1990s and bought a small patch in Savar. With the support of his armed guards, Rana seized a portion of the lot from a former business partner; for another adjoining parcel, he took control by forging the deed. Law enforcement stayed silent. As one of Rana's victims said, "The police were scared."

In 2006, the Ranas constructed a six-story compound to house garment factories, shops, and a bank. The edifice rose fast and cheap, without regard for zoning laws or safety codes. In 2011, Rana managed to secure a permit to build two additional floors—locals suspect due to discreet bribes, a common play in Savar. The town "grew quickly, and in an unplanned manner," a former local politician admitted. "There are so many buildings like Rana Plaza."

On the morning of April 23, 2013, workers at the five manufacturers based in Rana Plaza were busily sewing when an explosion rattled the building and split a second-floor wall apart like a fault line. "The crack was so huge I could put my hand in it," Shila Begum, a short, stout young woman who worked as a sewing machine operator for Ether Tex Ltd., on the fifth floor, remembered five years later.

Terrified employees poured into the street. Management called in an engineer to inspect the damage. He wanted to condemn the building immediately. Sohel Rana, who was at Rana Plaza meeting with reporters, refused. "The plaster on the wall is broken, nothing more," he reportedly said. "It is not a problem." Everyone was sent home but ordered to return the next morning.

Around eight a.m. the following day—a Tuesday—Mahmudul Hassan Hridoy heard a knock on his front door. It was his boss and neighbor, reminding him that they were expected at the factory. Hridoy, a good-natured twenty-seven-year-old, was in fine form: he had married his pregnant sweetheart of three years that weekend, and two weeks earlier, he had quit his poorly paid job as a nursery-school teacher for the far more fruitful position of quality inspector for New Wave Style Ltd., a fashion supplier at Rana Plaza. As he was good at math, management assured him he would move up the ladder quickly. "That's why I joined Rana Plaza," he told me over lunch at the KFC in Savar in 2018.

He listened to his boss and went to work, as did everybody else, including Shila Begum. "I was really in a panic," she recalled as we spoke on a Savar sidewalk under the burning noontime sun. They all showed up, she said, because if they didn't, they feared they would not be paid at the end of the month. Bangladesh's minimum wage was $38 a month at exchange rates then—or one-third of a living wage, which economists calculate is the amount needed to cover essential needs such as housing, food, and clothing. (In January 2019, it was raised to $95 a month, which is still half a living wage.)

"I was minding my own business, making blue jeans like you are wearing, for a French brand, when the power went out," she told me. "A couple of minutes later, the generators started." As the engines rumbled, the building began to quake.

And then, she said: "It went down."

She paused and looked at me. Her dark eyes were dim, as if someone had extinguished the light inside her.

"The concrete ceiling fell on my hand and my hair was caught in the sewing machine," she said. "After a lot of struggle, I untangled my hair, but I could not free my hand." Sixteen hours later, neighbors who had joined the hundreds of emergency responders at the site rescued her. "They showed up with iron rods and pipes and pried me out," she said. "They said my guts were all over the place. I passed out and came to my senses twenty-seven days later."

Hridoy was inspecting jeans on the seventh floor when everything went dark and silent. The generators started, he recalled, "and it felt like the floor under my feet was moving. Then it was disappearing." When he opened his eyes in the rubble, he realized he was pinned under a concrete pillar. As everything came into focus, he saw he was face-to-face with one of his good friends, Faisal, who worked on the second floor as a sewing machine operator. "I'm not sure how," Hridoy told me in a whisper. "I guess my floor dropped down to his." Faisal's skull was shattered. "And his brains were spilling out."

Hridoy began to cry. "I can't forget how his head exploded in front of me," he said, sobbing. "Those memories still haunt me."

With 1,134 dead and 2,500 injured, Rana Plaza was the deadliest garment factory accident in modern history.

"I lost all of my friends," Begum said. "Many were never found."

AT FIVE A.M. STOCKHOLM TIME, the telephone of Helena Helmersson, head of sustainability for H&M, rang, waking her up. On the line was her point person in Savar, who recounted the horror of the Rana Plaza collapse and assured her that H&M did not officially produce any of its clothes there. That said, she was warned, H&M's contractors *could* have subcontracted to one of the Rana Plaza workshops; there was no way to know until the investigation was completed. Given that H&M was Bangladesh's largest apparel exporter, even if the brand was in the clear, it might still be hammered by labor rights groups and consumers as a symbol of all that was wrong with offshoring: the lack of oversight and safety enforcement; the human rights violations; convoluted and untraceable supply chains. Two hours later, at seven a.m. sharp, Helmersson met with Karl-Johan Persson, chief executive of H&M, to carefully craft the company's response.

"None of the textile factories located in the building produced for H&M," it stated. "It is important to remember that this disaster is an infrastructure problem in Bangladesh and not a problem specific to the textile industry. The fact that this wasn't any of our suppliers' factories doesn't mean that we are not engaging in the process of contributing to constructive solutions."

Most brands, however, remained mum.

"No one stepped forward after Rana Plaza to acknowledge they were manufacturing there," Foxvog told me.

To root out which brands did produce there, teams of researchers spent months combing through the debris looking for labels and scanning import databases and factory websites in search of sourcing information. "A real triangulation," she said.

When confronted with the evidence, Walmart claimed its orders had been subcontracted to Rana Plaza without company authorization, Carrefour of France denied producing there, and JCPenney and Lee Cooper/ Iconix said nothing. Once it became clear that more than a dozen US and European brands' clothes had indeed been made there, most dodged calls for compensation to victims' families and survivors, and as there were no worker rights agreements in place, there was no obligation to pay.

This time, after an onslaught of highly embarrassing media coverage, brands were in a sweat. The double whammy of Tazreen and Rana Plaza was too much. They had to do *something*. Then, they remembered the initiative they had outright ignored for two years: the Bangladesh Fire and Building Safety Agreement.

Within six weeks, forty-three companies, including Primark, Inditex,

Abercrombie & Fitch, Benetton, and H&M had signed the charter, renamed the Accord on Fire and Building Safety; by October, there were two hundred members, among them Fast Retailing (parent company of Uniqlo) and American Eagle.

A slew of other brands—mostly American—balked, citing liability concerns. In July, Walmart announced the enactment of the Alliance for Bangladesh Worker Safety, a similar agreement. Among its signatories: Gap, Target, Hudson's Bay Company (owners of Saks Fifth Avenue and Lord & Taylor), and VF Corporation (owners of Lee Jeans, Wrangler, The North Face, and Timberland). But the Alliance was *not* legally binding, and NGOs saw it as less efficient, and sincere, than the Accord. It was "smaller, and worked with smaller factories," Foxvog explained. And it was voluntary—a system that had been proven by decades of fires and collapses as useless.

News stories on Rana Plaza were blunt and inescapable. The awareness campaigns that followed were vocal. Yet Americans didn't change their apparel shopping habits. In 2013, they spent $340 billion on fashion—more than twice what they forked out for new cars. Much of it was produced in Bangladesh, some of it by Rana Plaza workers in the days leading up to the collapse.

IN APRIL 2018, I traveled to Bangladesh to see if conditions in garment factories had improved since Rana Plaza.

My conclusion: yes. And no.

First: yes.

In five years, 97,000 safety infractions, such as locked doors, nonexistent fire exits, and dangerous wiring had been corrected in 1,600 factories. The government had shut down 900 that did not meet compliance standards.

NYU's Stern Center for Business and Human Rights published a study a week before my visit that stated there were still $1.2 billion worth of repairs needed to "remediate remaining dangerous conditions" in Bangladeshi garment factories and called for a task force to oversee the identification and rectification of those infractions. But the Accord had just been renewed for three years, and there were signs of progress.

I saw the improvements firsthand at an Accord-conforming factory called Rizvi Fashions Ltd. A six-story, mauve-stucco building on the road to Savar, it was constructed in 2014 by Shakil Rizvi, the former president of the Dhaka stock exchange. Inside, more than 2,000 workers manned the 1,450 machines and produced 2 to 2.5 million garments every month.

Rizvi Fashions was spanking clean and secure. Teams of men and women, most under thirty, were turning out slate-gray cotton briefs for Primark, bubblegum-pink terry shorts with ruffle hems for Deltex Organic, and white cotton undershirts for Fruit of the Loom. Everyone, from sewing machinists to quality inspectors, donned a face mask to block the inhalation of fibers. Veiled women pushed wide, mop-like sweepers—homemade Swiffers—up and down the aisles nonstop to keep the place spick-and-span. The fluorescent lights were bright, and wall-long windows allowed the sun to pour in. Huge ventilation fans cooled and circulated the air, and though it was already pushing 100 degrees outside, on the production floor the temperature was bearable.

When the factory was built, it adhered to Bangladeshi regulations at the time, which meant wiring was exposed and dangling from the ceilings, and there was nil in the way of fire safety. Then Accord inspectors came around, and things changed. Now the wiring is insulated and encased. There are sophisticated fire alarms; emergency floodlights; shelves of fire extinguishers, axes, and helmets; monthly fire drills; and fire safety teams, with leaders dressed in Day-Glo yellow vests so workers can find them easily. One-fifth of the workforce undergoes training with the local fire department to learn the basics of first aid, firefighting, and rescue. "It's an ongoing process," Anwar Hossain, manager for compliance, told me, as we crossed the plant floor.

There were other perks that sweatshops lacked: a company-subsidized canteen where workers can buy sandwiches and drinks at cost; an infirmary; a daycare center. When I looked in, a toddler and a seven-year-old girl—daughters of two different workers—were playing together. The factory is closed at night and every Friday, the Muslim holy day.

Hossain then guided me across the parking lot to the "pump room," an independent building with a sophisticated hydraulic system connected to a private 150,000-gallon underground reservoir. He and his team were mighty proud of that installation.

The Accord-demanded precautions paid off: Rizvi has never had a fire.

BUT THAT DIDN'T mean there weren't still sweatshops in Bangladesh.

"I've seen much worse," my guide told me, as we pulled up to one in Dhaka. "We just drove past several that are much worse."

Set on a blighted backstreet, the factory we were to see was built in the early 2000s, but it was so dilapidated, it looked twice as old. Four uniformed security guards—older men who seemed not terribly menacing—sat at a small table at the entrance. *As-salamu alaykum*—or "Peace be unto you"—one said to me, raising his hand to his face and nodding slightly.

We made our way up the building's lone staircase to the first-floor reception area. The filth splattered on the walls was unfathomable.

We were invited to tea with the boss and his team. I was offered an old farm chair, its paint peeling. My translator took a seat on a plastic stool. The boss settled in behind his Formica-topped desk on a well-worn pink Naugahyde swivel chair, the stuffing spilling out the split side seams. It was as if the entire office suite had been found in the street.

A wisp of a girl with wide-set eyes, enrobed in a magenta sari and tangerine cotton veil, glided by silently.

"She's young," I remarked quietly to my translator.

"Fifteen maybe," he responded just as quietly.

After five minutes of chitchat between our facilitator and the boss, we were given a tour. The stairwell—the only way out—was encumbered by big cartons of product bound for Russia. On the wall of the landing, large signs titled BUSINESS SOCIAL COMPLIANCE INITIATIVE listed, in Bangla and in English, all the safety and labor rules the factory supposedly followed. The facilitator whispered: "Those are fake."

Next to the doorway, red fire buckets brimmed with trash; the big black plastic cisterns were cracked and half empty. We entered the sewing room— a long, dark space with weak lighting. More than a hundred workers, very old to very young—some visibly in their early teens—sat behind beat-up machines, sewing clothes at a frantic pace. Like the girl I had seen earlier, the women were dressed in vibrant-colored saris and headscarves. Everyone was barefoot. No one wore surgical face masks, despite the SAFETY FIRST and WEAR A MASK signs tacked on the walls. Wiring was exposed. Windows, many with broken panes, were barred. Fabric scraps, stray threads, and heaps of finished clothes littered the floors. It was unbearably hot—more than 100 degrees Fahrenheit. Small fans blew the dust around but didn't cool much off.

There were several more loft-like floors, each dedicated to a different task. In the finishing room, tables were piled high with black jeans. Young women and men, shoeless and standing on scraps of cardboard, inspected the pants, clipped off threads, and mended flaws. In the cutting room, workers drew and cut paper patterns by hand, rather than using the computer-run systems that are the norm in aboveboard factories today. Bolts of cloth lay about on the floor, willy-nilly—more obstacles to block escape if the place catches fire. The fabric cutter, a young man of maybe twenty and thin to the point of appearing malnourished, wielded an old electric hand-jigsaw, pulling it quickly with his right hand through five-inch-thick wads of denim toward his left, which was simply clad in a metal mesh glove—as if that was going to stop the machine from slicing off his fingers. In modern plants, cloth cutting is automated.

Once we were back on the street, our facilitator told us the factory owners were a married couple who practiced ophthalmology, and they were never on-site. He said Walmart and Lidl had been produced there, but he didn't know if these were contracted or subcontracted jobs.

We thanked him for his help, got in our tired, seatbelt-less Toyota Uber, and wheeled back into the traffic. As we sat, and sat, with auto-rickshaws and scooters and buses beeping and honking nonstop, my translator googled the factory on his phone to see what more we could learn. A website popped up, with photographs on the homepage of smiling workers in a bright, clean facility, and the administrative team sitting in neat, new office cubicles.

According to the introduction, the firm was founded nine years earlier and had seven hundred workers and six hundred machines. It produced clothes "100 percent for export," and claimed its clients included Camaïeu of France, Roadrunner of Canada, and CJ Apparel of Great Britain. I wondered if any of its customers—like the Russians whose address was on the cartons in the staircase—had ever stepped into that factory.

TAZREEN'S OWNER, Delwar Hossain, was eventually arrested and charged with culpable homicide, meaning he knew his disregard for the factory's safety was likely to cause death. The trial, initiated in 2015, dragged on, as the prosecution failed to produce witnesses, and by the sixth anniversary, in November 2018, it had not been concluded.

In 2016, Sohel Rana and seventeen others, including his parents, Rana Plaza's engineer, Savar's mayor, three government inspectors, and its town planner, were charged with various crimes, including homicide. A year later, Rana was sentenced to three years in prison for not revealing the true amount of his personal fortune to the Anti-Corruption Commission. The trial for murder and other charges remained on hold, due to appeals to a higher court.

While not efficient, both cases were unprecedented in the Bangladeshi garment industry.

Back in 1998, only about 15 percent of company codes of conduct included freedom of association and collective bargaining, Gearhart told me. Now it's a given. Brands are publishing supplier lists—unthinkable even five years ago—and when a worker dies, they pay compensation.

As with every other improvement in Bangladeshi apparel manufacturing, the impetus was Rana Plaza. In the wake of the disaster, brands suddenly felt compelled—either by genuine guilt or, more likely, fear of reputational damage—to pony up something for the loss of life or limb. "But it's not called 'compensation,'" Foxvog made clear. "It's called the 'Rana Plaza Arrangement.'" Hammered out by Clean Clothes Campaign (CCC), the $30 million endowment was underwritten by brands to help ease the burden for families of the deceased as well as those who sustained catastrophic wounds. But access to those funds has been complicated.

Shila Begum's injuries were severe: she must wear a medical corset and a brace on her right forearm. "My kidneys were crushed and damaged," she told me. "I can barely use my right hand." She received no government compensation and "nothing from the brands," she said. Because of her disabilities, she cannot work, and, widowed, she has no other financial support. This forced her to pull her fourteen-year-old daughter out of school; while education is officially free, she could no longer afford incidentals such as books and lunch. For daily expenses, she confessed, "I have to beg my family for money." She began to weep. "When I come here," she said, looking at the vacant Rana Plaza lot before us, "I don't want to live anymore."

Hridoy, the newlywed, walks with a crutch and suffers from terrible headaches. Sometimes he pulls his hair out as he sleeps. During his long recovery, his wife left him and had an abortion. "Rana Plaza ruined my life," he said, wiping tears from his eyes.

Nonetheless, he has done his best to rebuild it. Hridoy was the only survivor of the dozens I met in Savar who managed to wrangle money from the Rana Plaza Arrangement trust fund. He used it to open a small pharmacy. He also founded the Savar Rana Plaza Survivors Association, a support group of three hundred members who gather informally once a month in his shop to help shore each other up. Sometimes, such camaraderie isn't enough: in 2015 and 2016, two association members committed suicide by hanging themselves in their living rooms.

There is still government opposition to freedom of association and unionization. Sexual and physical abuse in factories remain rampant. Workers still do not receive severance when manufacturers go belly-up.

They also continue to be at the mercy of the brands—even if the work is contracted. Terrorism has been a constant threat in Bangladesh since the summer night in 2016 when Islamic militant gunmen stormed a chic Dhaka café frequented by ex-pats and foreigners, took forty hostages, and killed twenty. Five militants, two police, and two café employees were killed when law enforcement staged a rescue. Fashion reps immediately canceled scheduled trips and withdrew staff from the country, alarming workers as well as factory owners. "What will happen to us if the buyers are not going to come to Bangladesh?" one seamstress, a mother of two, wailed. Western chain hotels fortified their compounds with cement barriers, metal detectors, x-ray machines, and guards wielding bomb-detecting wands. The industry rebounded.

And then there is the endless tussle over paying workers a living wage. In 2016, they staged protests and demanded immediate raises. Owners and the government retaliated with force: fifty-five factories shut down for a week; fifteen hundred workers were fired; thirty-five were thrown in jail, twenty-four without bail. "That was dark for us," Gearhart recalled. "Never have we had that many workers arrested or held that long or denied bail."

All this because "the industry is *obsessed* with quarterly returns," Mark Anner, director of Penn State's Center for Global Workers' Rights, told me. "How do you develop a long-term vision if every three months shareholders demand more profits or threaten to pull out? How does this trickle-down affect workers?" Or Bangladesh as a whole? "If the workers made a decent wage, the economy would grow, because they could afford to buy lunch and get a haircut," Gearhart said. "Where's the investment in workers today? Are they really just cogs in the wheel?"

# THREE

# Dirty Laundry

YOU ARE PROBABLY wearing jeans as you read this. If you're not, chances are you wore them yesterday. Or you will tomorrow. At any given moment, anthropologists believe, half the world's population is sporting jeans. Five *billion* pairs are produced annually. The average American owns seven— one for each day of the week—and buys four new pairs every year. "I wish I had invented blue jeans," the French couturier Yves Saint Laurent confessed. "They have expression, modesty, sex appeal, simplicity—all I hope for my clothes."

Barring basics such as underwear and socks, blue jeans are the most popular garment ever. They are what many of the Rana Plaza workers were sewing or inspecting when the building came crashing down. They were the backbone of American textile and garment manufacturing, until Levi's off-shored those jobs. All of them. They are hyper-polluting—in their creation, and in their afterlife.

Jeans embody all that is good, bad, and awry in fashion.

TRUE BLUE JEANS are made of cotton, one of humankind's oldest crops—
it is thought to have been domesticated in 3500 BC (though some archeologists
believe that could date to the sixth millennium BC). The ancient Greek historian Herodotus described cotton as "a wool exceeding in beauty and goodness that of sheep." When the Macedonian king Alexander the Great invaded
India with his army in 327–326 BC, he had his troops use cotton, which they
called "vegetable wool," for their bedding and saddle pads. In 63 BC, the
Roman official P. Lentulus Spinther had sun-blocking cotton awnings installed on the theater for the Apollinarian games, and two decades later, Caesar tented the Forum and the road that linked the city's palace and capitol
with cotton tarps.

About sixty billion pounds (or 121.4 million bales) of cotton are grown
each year on roughly eighty-two million acres (or 33.4 million hectares) of
land across more than one hundred countries—India is the largest producer;
China is a close second, and the United States is third. You can find cotton
in fishnets, coffee filters, book bindings, bandages, disposable diapers,
x-rays—even banknotes: America's paper currency is 75 percent cotton and
25 percent linen. Cotton's most common use, however, is for clothing: 60
percent of all women's apparel contains it; for men, it's 75 percent. For blue
jeans, it's 100 percent.

For all the comfort and shelter it has given, cotton is one of our most man-
handled plants. It has been spliced, and sprayed, and souped up until, like a
*Playboy* centerfold, it barely resembles its original self.

Nonorganic cotton—known in the business as "conventional" cotton—
is among agriculture's dirtiest crops. One-fifth of insecticides—and more
than 10 percent of all pesticides—are devoted to the protection of conventional cotton, though it is grown on only 2.5 percent of the world's arable
land. The World Health Organization has classed eight out of ten of America's most popular cotton pesticides as "hazardous."

Conventional cotton is also extremely thirsty: to grow one kilo requires, on average, 2,600 gallons (or 10,000 liters) of water. Processing it swallows even more: approximately 5,000 gallons for one T-shirt and a pair of jeans. If fashion production maintains its current pace, the demand for water will surpass the world's supply by 40 percent by 2030.

Much lore surrounds the naissance of denim and jeans. Some historians claim the durable indigo-dyed cotton twill originated in the southern French town of Nîmes—or *de Nîmes*, as the French say. Some say Christopher Columbus of Genoa—or "Gênes," in French—used the fabric for his sails and that Genovese sailors wore blue cotton pants.

Fashion scholars now believe denim as we know it today was developed by the Manchester, New Hampshire, textile mills during the nineteenth century and named "jean." The fabric's construction has always been simple: the exterior side, known as the warp, is made of two or three yarns, usually dark blue, woven together as one; the underside, called the weft, is a single yarn in white or another pale shade, giving the fabric a 3-D–like effect. There is a higher quality, called "selvedge." It is woven on older, narrower shuttle looms (half the size of the common ones, which are roughly sixty inches wide) with a continuous weft thread. Selvedge denim is tighter and more substantial, and its outer edges are woven to avoid fraying—they have a "self-edge." Selvedge jeans aficionados show off the seam edges, as if an ensign, by turning up their cuffs.

Originally, the New Hampshire–spun yarn was dyed with natural indigo farmed in the American South. Indigo is one of humanity's oldest natural dyes, extracted from the leaves of the *Indigofera* plant. It was first cultivated in the United States in the mid-eighteenth century by slaves of Eliza Lucas Pinckney, an Antiguan of English descent. She sowed seeds sent by her father, the lieutenant governor of Antigua, on her family's South Carolina plantations. Like cotton, indigo was farmed by slaves, and it soon became one of the Southern colony's most lucrative crops.

Denim remained a niche textile until the early 1870s, when a Reno, Nevada, tailor named Jacob Davis wrote to his fabric supplier, Levi Strauss, a

Bavarian-born immigrant with a successful dry goods business in San Francisco, and asked for help mass-producing his most recent design: workpants with metal rivets at key stress points. The pants were such a hit among miners, farmers, and laborers, Davis claimed, "I cannot make them fast enough."

If Strauss would cover the hefty $68 patenting fee, Davis proposed, the two men could be business partners, manufacturing the riveted trousers in San Francisco. Davis included with his letter two versions of the pants—one in an ecru-colored canvas known as "duck," and another in denim. Strauss liked what he saw and applied for the patent. On May 20, 1873, it was granted. Davis settled in San Francisco and within weeks was overseeing the first run of Levi Strauss & Co. jeans. Made of denim from Amoskeag Manufacturing Company, a mill in New Hampshire, the pants were sewed by fifty seamstresses in the duo's new factory on Market Street. Today, Levi Strauss & Co. design and sell the majority of jeans. It is one of the most successful apparel brands, ever.

The world's oldest surviving Levis can be found in the Vault, the company's archives at its headquarters, Levi's Plaza, an early-1980s brown brick and dark glass office complex across from the Embarcadero in San Francisco. The Vault is watched over by company historian Tracey Panek, a kind, keen middle-aged woman who calls to mind an elementary-school librarian.

On a sunny fall Friday in 2017, Panek—dressed in black Levi's jeans, a well-worn blue denim jacket, and a bright red turtleneck, perhaps a nod to the company's signature scarlet tab—welcomed me into the Vault to view the best of Levi's vintage collection. We started with a pair from 1879. Panek tugged on her white cotton gloves, pulled open a drawer in one of three fireproof safes, and gingerly lifted the pants out to place them on a long archival table.

They were short, wide, and a faded chalky blue, with century-old dirt so embedded on the thighs, it would never, ever come out in the wash. Panek explained that, because jeans weren't cheap, miners would mend and share them until they had no life left. "I call them the first sustainable garment, because you could patch them up and pass them on," she said.

In 1890, Strauss and Davis introduced a new jean silhouette called the "501," named for the lot number. Panek showed me a pair from that era. Made of a denim known as "XX," for extra strong, they had buttons for suspenders (instead of belt loops), a button fly, and four pockets—three in the front, including a small inset for a pocket watch, and one on the back; the fifth, on the derriere, was added in 1901. Since then the 501's silhouette has gone unchanged.

The profits from his blue jeans empire made Strauss one of California's wealthiest citizens. He wore it well: standing about five feet six and gently rotund, he was always impeccably turned out in a dark broadcloth suit with a waistcoat, silk tie, and top hat. He did not don the jeans his company produced. He never married either, claiming, "my entire life is my business." When he died in 1902, at seventy-three, he left much of his nearly $1.667 million estate to local charities, and the company in the hands of his four nephews. One of them, Sigmund Stern, and Stern's son-in-law Walter Haas, took charge. They modernized the jeans' design, replacing the button fly with zippers, the suspender buttons with belt loops. And while they continued to buy their denim from Amoskeag Mills, they added a new source in the South: the Cone brothers' White Oak Cotton Mills, in Greensboro, North Carolina.

Cone Mills was founded in the mid-1890s by Moses and Ceasar [sic] Cone, two brothers who, like Strauss, were of Bavarian stock. Hearing of Strauss's success out West, they spotted a bankable business opportunity and transformed a defunct steel mill into a denim factory called Proximity Manufacturing, since it sat close to cotton fields and gins. In 1905, they inaugurated a second facility, White Oak Cotton Mills, named for a majestic two-hundred-year-old specimen that stood nearby.

Much of the denim produced there was dyed with synthetic indigo, developed by the German chemist Adolf von Baeyer and commercialized by BASF (Badische Anilin und Soda Fabrik) in 1897. Unlike natural indigo, synthetic wasn't seasonable, or vulnerable to blight or weather destruction. Yes, it was made of a host of chemicals—several we now know are harmful to the environment. But it was consistent, and cheaper. This meant mills like Cone could weave and dye denim twelve months a year. By 1914, the natural indigo business had been annihilated, never to recover. White Oak eventually became the largest denim producer in the world, and Moses was known as the Denim King.

In 1915, the Levi Strauss nephews met with the Cones to discuss sourcing denim from White Oak. The two sides came to swift agreement, sealed with what has gone down in history books as "the Golden Handshake." From then on, Cone exclusively supplied Levi Strauss with denim for its 501 jeans. And blue jeans' popularity steadily grew, until they received an unexpected bump in the 1970s—from all places, Seventh Avenue.

With the women's liberation movement and the popularity of more casual dress, New York's fashion designers dreamed up a new fashion category: designer jeans. With slim-cut legs and seats that cupped the rear, designer jeans embodied the epoque's hedonism and Madison Avenue swagger. "Jeans are sex," Calvin Klein said. "The tighter they are, the better they sell."

To hammer home his point, in 1980, Klein cast fifteen-year-old actress-model Brooke Shields for his jeans commercial. "You want to know what comes between me and my Calvins?" she purred in her childlike voice, as she sat spread-eagle in a pair of his jeans and a taupe blouse. "Nothing." The ad was so provocative, the New York affiliates of ABC and CBS promptly banned it. But it had already worked its spell: Klein sold four hundred thousand pairs the week following the ad's debut, then two million a month after that. Jean sales rocketed to record heights: more than half a billion were purchased in 1981 alone.

UNTIL THE 1970S, a good many jeans sold had been made of stiff, shrink-to-fit—or "unsanforized"—denim. (Preshrunk cotton—known as "sanforized"—has existed since the early 1930s, but was not adopted by the denim industry until the 1960s, when prewashing became a common practice.) One would buy unsanforized jeans a size or two larger, then, to get them to fit, either wash them or—ideally—put them on and sit in a bathtub full of water. Really.

To soften them, you simply had to wear them. A lot. It took a good six months to properly break in jeans. After a couple of years—*years*—the hems and pocket edges might start to fray, or a knee would split open. The fabric faded to a powdery blue with some whiskering—the sunburst-like streaks that radiate from the fly. Time and dedication were required to push your jeans to peak fabulousness.

That is, until the popularization of stonewashing in the 1980s. Unsanforized jeans were thrown into industrial washers with pumice stones and tumbled until the denim was sufficiently abraded. (The L.A.-based casualwear company Guess famously had a system that stonewashed jeans for seven hours—a marathon now considered an environmental horror.) Sometimes jeans were further distressed with acid, sandpaper, rasps, and files to mimic the previously hard-won wear and tear. The entire operation was christened "finishing" and conducted in "washhouses," sprawling facilities that now process thousands of jeans a day.

Some washhouses—especially those in Los Angeles, America's jeans-finishing center—are highly technical and follow strict worker safety and environmental norms. But a lot do not, as I saw in Ho Chi Minh City on a steamy April morning in 2018.

Vietnam's textile and apparel industries are both old and young: for centuries, local women spun and wove silk into exquisite cloth for home and dress, but the Arkwright-like factories that spit out fabric and garments by the ton weren't prevalent there until the mid-twentieth century. Apparel and textile manufacturing amounted to a small slice of Vietnam's GDP for decades: when I visited the country in early 1993 and drove from Hanoi south to Da Nang, I spotted a few factories, but it was clear—as evinced by the emerald lawn of rice paddies across much of the land—that the nation's economy was greatly agrarian.

Trade agreements and globalization changed that landscape. By 2018, there were roughly six thousand textile and garment production companies in Vietnam, employing 2.5 million workers, and accounting for about 16 percent of the country's exports and more than $30 billion in revenue. Experts believe that last figure will jump to $50 billion by 2020.

Much of the work is jeans finishing. In 2012, jeans production turnover in Vietnam was $600 million; by 2021, it is expected to double.

On the industrial outskirts of Ho Chi Minh, a local denim expert and I rolled up to a run-down, warehouse-like plant behind an unassailable gate. Inside, where about two hundred young Vietnamese labor, the fluorescent lighting was poor and it was 100 degrees, easy. Large fans whirred to try to cool the room. It didn't work.

Pristine midnight-blue jeans were piled high on metal tables and dollies. Young men in butter-yellow T-shirts, trousers—usually jeans—and knee-high rubber boots took them and stuffed them into two dozen monster-sized washing machines. Other young, booted men pulled wads of sopping jeans out of the machines. An inch of navy-blue water stood on the floor. The men did not wear gloves, and their hands were stained black.

Some of the machines were older types that require twenty liters, or five gallons, of water to wash one kilogram—three pairs—of jeans. Others were less piggy, using only five liters—or a bit more than one gallon—of water per kilo of jeans. Manufacturers "know how wasteful this is," my guide told me.

And costly: wastewater must be processed; no longer can Vietnamese washhouses dump straight into the waterways—thankfully. I saw a canal on the other side of town where jean-washhouse waste had been discharged for years. The water now resembled tar, and its stench made me want to vomit.

But it's hard to get factory owners to change their methods.

"Their business is about washing, not about worrying about the planet," a jeans expert told me.

The wet jeans were heaved into enormous crates and wheeled to another room, where they were thrown into massive dryers. Some of the jeans were treated with chemicals and baked in a giant oven to replicate whiskering. This is called the "dry process."

In the distressing room, young men and women in sky-blue T-shirts—each department had its designated color—were sanding jean knees and thighs by hand, like a carpenter works on wood. Some wore medical masks to prevent inhalation of denim dust, but most did not.

The verve with which they attacked their assignment was alarming: each pair went from virgin to wrecked in under a minute. The workers' focus was intense—they didn't speak or notice anything happening around them. One slipup and their pay would be docked. At the time I visited, sanders processed at least four hundred pairs of jeans a day, six days a week, not including overtime.

And that was the *hand* distressers. The machine distressers worked even faster. I watched one woman tackle cutoff shorts with what looked like an oversized dental drill that emitted a scream so high-pitched it could crack crystal. She ground the front and back pockets and hems of those shorts to a fashionably holey state in ten seconds. Six pairs a minute. All day long. She was unmasked in a room where it was hard not to sneeze.

This all apparently compared well to the washhouses of Xintang, the town in Guangdong Province, China, that claims to be the "jeans capital of the world." Each year, 200,000 garment workers in Xintang's 3,000 factories and workshops produce 300 million pairs of jeans—800,000 pairs *a day*. The local water treatment plant closed years ago, leaving factories to dump dye waste directly into the East River, a tributary of the Pearl River. It turned opaque; aquatic life could no longer survive. Greenpeace has reported that the riverbed contains high levels of lead, copper, and cadmium. Xintang's streets are dusted blue. And many garment workers have reportedly suffered from skin rashes, infertility, and lung infections.

IT DIDN'T HAVE to be this way, as cotton expert Sally Fox explained to me. We were sitting at a simple wooden table in Fox's double-wide trailer home on Viriditas Farm, her 130-acre stead in the Capay Valley, northwest of Sacramento, on an early autumn morning. The living room was lined with stacks of cardboard cartons and rows of oak filing cabinets stuffed with dossiers on cotton: studies, orders, swatches. The windows were open. The quiet was broken only by her merino sheep bleating in the pasture, a rooster crowing in the barnyard, and the north wind rippling through the shade tree out front.

Five foot seven, with a rime-white bob and an honest smile, Fox was dressed in a water-blue chambray tunic and caramel denim jeans. Her face was appropriately lined for her sixty-one years and devoid of makeup; her turquoise eyes were clear.

I had come to see Fox because she is considered by many in the industry to be the mother of modern organic cotton. A native Northern Californian, she bought her first spindle at twelve with her babysitting earnings and started spinning wool, cotton—anything she could find. While in the Peace Corps in Gambia in 1979 and 1980, she helped develop natural ways to fight pests. For the last forty years or so, she has been breeding and farming colored organic cotton in Arizona and California.

Colored cotton has been around since *Gossypium barbadense* was first domesticated and has existed in an array of hues: brown, tan, green, blue. The Chinese grew a pale-yellow variety used for cloth called nankeen that was popular in the American colonies, Fox told me during our chat in her trailer. "Everyone wanted nankeen gold trousers."

After she finished her master's, she went to work for independent plant breeder Robert Dennett, near Davis, California. While cleaning the greenhouse one day, she opened a drawer and found a bag of brown cotton. The fiber was short, weak, and rough, she recalled. But she was charmed by it and thought, *If it could be spun, people would want it, because you don't have to dye it.*

She ordered seeds from the USDA and planted them in pots in Dennett's greenhouse.

Fox was so pleased at how they turned out, she planted a quarter acre of land near Bakersfield, the capital of California cotton farming. "The next year," she said, "I rented an acre, then five, then eleven, and on and on." She discovered the tannins that made the cotton colored also made it naturally disease and insect resistant, so she farmed organically, back when "no one was doing organic cotton," she said. "No one."

She launched Natural Cotton Colours Inc., a company to sell her colored cotton, which she named FoxFibre, and began collaborating with independent fashion designers as well as landing production contracts. One was with Levi's. Fox's cotton was used to develop a caramel-colored denim, and her friend Dan DiSanto, then a Levi's designer, sourced it for a new line. The jeans Fox was wearing the morning we met at her farm were made of that denim.

Levi's and Fox negotiated a three-year deal. She would supply colored cotton seeds to farmers in West Texas, who would grow the cotton, spin it, and weave it into denim at a mill the farmers owned together as a co-op. Then Levi's would buy the finished denim and turn it into clothing. The first year, the farmers planted one hundred acres' worth of Fox's seed, and Levi's bought the denim. The second, the order was upped to a thousand acres. And the third year, it was three thousand acres. "The farmers made so much money," Fox remembered. "They were really happy."

As was Levi's. The jeans and jackets were "wildly popular," she said.

When Fox reached the one-thousand-acre mark, Levi's head (and Strauss's great-great-grandnephew) Bob Haas told her, "This could change the world. If you could get a hundred thousand acres' worth of seed, I can do it, I can make it happen."

"It will take me two years," she responded, "but I could do it."

"Please do it," he said.

This time, Fox and Haas didn't draw up a contract. "I just made it my goal because I wanted so much to be part of the reduction of this enormous environmental disaster," she told me. "I bet my business on that. I got all the seed, and I paid a million dollars to do it and prepared to plant the seeds for one hundred thousand acres."

As Fox was busy on the project, Levi's plunged into a management crisis.

In 1996, when the brand was reporting a record-breaking $7.1 billion a year in sales—more than Nike—Haas initiated a leveraged buyout. The deal gave voting rights to a clutch of relatives but incurred $3.3 billion of debt, which was listed on the stock market, putting the company in a precarious financial situation.

Turns out, 1996 was Levi's revenue apex: the following years, sales tumbled unexpectedly, as the company lost market share to fiercely competitive start-ups. Heads rolled.

Fox went to meet the new executive team in San Francisco—Haas was still CEO—and, she remembered, "the vice president went on this rant about how he hated brown and green. 'When do you ever see a green car?' he said."

Levi's canceled the cotton order.

Fox's company, Natural Cotton Colours, filed for Chapter 11 bankruptcy.

She gazed silently out the window of her double-wide, across the Northern California prairie.

"I lost everything."


FOR MUCH of the twentieth century, Levi's was known in the apparel industry as a company with a conscience. In part, that was because the family that ran it—Strauss's descendants, the Haases—were devout Jews who carried on his commitment to charity and philanthropy, and in part because the company was headquartered in politically liberal San Francisco. In the

1970s, Levi's chief executive Walter Haas Jr. brought in a religious ethicist to advise him on how to adopt more responsible business practices. The company famously pulled its operations out of Indonesia in the mid-'70s because of the country's widespread corruption and chose not to enter the South African market because of the government's racist apartheid policies. In the early 1980s, Levi's became one of the first American corporations to address the AIDS epidemic, drawing up corporate standards to support HIV-positive employees.

When Walter Jr.'s son, Bob—a former Peace Corps volunteer in the Ivory Coast—took the helm in 1984, he upheld that ethos. "It is more than paternalism, really," he said in 1990. "A company's values—what it stands for, what its people believe in—are crucial to its competitive success."

But now the business was sinking. The teen market had moved on to hipper brands such as Gap and Tommy Hilfiger, and the fashion consumer to hot new specialty labels. To make up for the precipitous revenue drop, Levi's started selling at discounters like Kohl's, further denting its rep. "Levi's lost sight of who they are," Morgan Stanley Dean Witter apparel analyst Josie Esquivel said at the time. It had "lost its cachet."

And its moral compass. As sales evaporated, so did the company's values. In 1997—three years after NAFTA—Levi's announced it would close fourteen plants across the US and Europe, citing high labor costs.

The hit to employees, and the towns where Levi's had operated for decades, was brutal, as I heard from Annabelle Nichols, a straight-backed southern woman who, when I met her in 2016, was seventy-four years old. Nichols had spent the first four decades of her garment-manufacturing career at Levi's Cherry Street plant in Knoxville, Tennessee.

The Knoxville facility was Levi's largest in North America. It opened in 1953, in a former tobacco warehouse, and was divided into four sections, each about the size of a football field, separated by cement-block walls and steel sliding doors. Almost all the employees—95 percent—were women. They worked from 7 a.m. to 3:30 p.m., Monday through Friday, from 6 a.m.

on Saturdays, making twenty thousand pairs of blue jeans a day, as well as khakis "and dress slacks," Nichols said. The remaining 5 percent of staff were men, and they were managers. There were no female managers; women who oversaw workers were dubbed "supervisors." They were paid less than, and reported to, the men. Nichols joined the workforce in 1961, at nineteen, and after seven years, she was promoted to supervisor.

On Monday, November 3, 1997—the day before election day—as the plant floor rattled with its usual Gatling-gun-like racket, an authoritative voice came over the public-address system and asked workers to turn off their machines.

Once the cavernous space was silent, the voice continued, more mournfully:

"We have some devastating news . . ."

The factory would cease operations by the end of the year.

Workers began to weep at their stations.

At the time, Levi's employed 37,500 people in more than fifty factories worldwide. Thirty-two of those facilities were in the US. Five were in Canada. Levi's announced it was going to close eleven, including Cherry Street, immediately. That amounted to nearly 6,400 production employees—1,800 of them in Knoxville—or 34 percent of Levi's North American workforce.

Levi's swore it wasn't outsourcing the work: a spokesman said the downsizing was due to consumers spending less on apparel, from 7 percent of one's income in the 1980s to 4 percent in the 1990s. The company offered generous severance packages and job counseling to what it called "dislocated workers." That year alone, Levi's cut some 43 percent of its global workforce.

Nothing really eased the pain of the layoffs. Nichols retired briefly, then returned to work, as a production manager for Omega Apparel, a military uniform manufacturer, in Smithville, Tennessee.

Not everyone was so lucky. "We lost a lot of good people," she told me. "Several passed away right after."

LEVI'S SALES CONTINUED to free-fall—revenue slid to $5.1 billion, a 28 percent drop in three years. The company announced more plant closings, to "give the company greater flexibility," said Levi's American division president John Ermatinger. But Bruce Raynor, secretary-treasurer of the Union of Needletrades, Industrial and Textile Employees, saw the strategy for what it truly was: "[Levi's] decision to join the race to the bottom."

To execute the closures—and captain a turnaround—Levi's hired fifty-two-year-old Philip Marineau, most recently the president and chief executive of Pepsi-Cola North America, as CEO. He was the first non–Strauss family member to helm the company. (Haas remained chairman.) Marineau said he'd use the same methods that he employed to sell Gatorade and Mountain Dew, since, he claimed, without irony, soft drinks "aren't dissimilar to the fashion business."

The plan was simple: Levi's "had to go from a company that was a self-manufacturer to a creator, marketer and distributor of apparel," he said. In other words, it was going to subcontract *all* production, which would in turn be subcontracted, and so on, and so on.

That meant shutting down the last Levi's-owned facilities, among them the "Mother Factory" on Valencia Street in San Francisco, opened in 1906, and a plant in Blue Ridge, Georgia, an Appalachian Mountain town of fourteen hundred, that had been in operation for forty-three years. Levi's had already laid off three hundred at Blue Ridge a year earlier. Now it was giving pink slips to the final four hundred.

These were blue-collar jobs. Most workers earned $8 to $14 an hour for such tasks as making belt loops, attaching rivets, and sewing zippers into jeans. Some took home as little as $20,000 a year. But Levi's made up for the low pay by being a generous member of the community—donating to the hospital, schools, nursing homes, the public library, and Little League teams. In 2001, it gave $10,000 to the fire department for a new communications system, and each Christmas, it sent small bags of grooming products or a bit of cash to senior citizens at the local health center. Over the years, the company contributed several thousand dollars toward the county's first mobile defibrillator and helped pay for a Jaws of Life hydraulic rescue tool for car crashes, curling irons for a cosmetology course, and field lights at the stadium. "They've just allowed us to have a lot of things we couldn't have had," the high school principal, Doug Davenport, said.

That was over. Blue Ridge became an emblem for the economic and social destruction that profit-driven short-term decisions taken in boardrooms—like offshoring—caused in small manufacturing towns across America.

The state opened an employment agency to help former Levi's employees find new jobs—not an easy task considering few had a high school diploma. Local kids weren't taking swimming lessons at the county rec center, because former factory families could no longer afford the $20 enrollment fee. Many moved away in search of work, shrinking school enrollment, forcing the school board to fire teachers. Revenue sank, which led to cuts in public services. "Money is going to be tight," said Fannin County's recreation director, Bernie Hodgkins. "It's going to devastate this little county, I feel."

Levi's took a beating in the national press for pulling out of Blue Ridge—but that didn't slow Marineau down. Quite the opposite. In all, he sacked 25,000 Levi's employees. He defended his actions in an interview with the *San Francisco Chronicle*: "From a justice standpoint, there's no reason to say that the person in San Antonio deserves that job versus the person in Pakistan."

In the midst of firing all those folks, in 2004, Marineau reportedly earned $6.3 million in salary, bonus, and long-term incentive payouts and was up for an additional $4 million bonus over the next two years. At the close of those two years—at the age of sixty—he stepped down. His pension was $1.2 million a year.

And Levi's sales were still tumbling: $4.19 billion—almost half of its peak a decade earlier.

AFTER HER BANKRUPTCY, Sally Fox regrouped. Through an "ag exchange," she traded her farm in Kern County for the one in Brooks where we were sitting that October morning. "Same amount of acreage, though this is much prettier," she said, gazing admiringly out the window at her rolling terrain. She arrived with a truck, a travel trailer—which became her home, until she upgraded to the double-wide in 2003—and her seeds. She kept breeding and growing small amounts of cotton, to keep the lines alive. She financed everything by selling items she produced—yarns, socks, sweaters—on her website. All is 100 percent organic.

Mother Nature created cotton as a perennial. During its first year, it "wants to grow and become a huge tree," Fox told me, "and in the second year it puts out its flower. If you want the plant to flower its first year, it has to be stressed, by not giving it enough water or fertilizer; low-fertile soils will cause it to flower. That's why cotton always *had* the reputation of being a poor-soil crop—it was your last-resort crop, before you put a lot of manure or cover crops on your field. If you didn't have any money to bring fertility back to your soil, you could put cotton in and earn some."

But in the 1980s, BASF developed a mepiquat chloride–based growth regulator called Pix that effectively turned cotton into an annual: when applied to the plants, it triggers flowering. As cotton farmers are paid by the yield rather than by the acre everywhere except in the European Union—where production, in Greece and on the Iberian Peninsula, is small—this scientific breakthrough upended the cotton business. Farmers began to irrigate and water their plants profusely to push growth, then apply Pix. "Suddenly you could go from one bale of production an acre to six bales," Fox said. By the 1990s, most conventional cotton farmers were using Pix.

To keep weeds down, the American multinational agrochemical and bio-technology company Monsanto introduced, in 1997, a commercial variety of genetically modified cotton called "Roundup Ready Cotton." The seeds were engineered to allow the plant to endure heavy spraying of Roundup, Monsanto's "broad spectrum" glyphosate-based herbicide. In effect, Roundup would kill everything *but* the cotton. To work efficiently, each needs the other, forcing farmers to buy both. Monsanto's competitors have come up with their versions, too. And conventional farmers have bought into it: in 2018, 94 percent of American cotton was genetically engineered. In Alabama, Arkansas, Louisiana, Mississippi, and Missouri, it was up to 99 percent of the cotton farmed; in Georgia, 100 percent.

Roundup is the world's most heavily used pesticide, and it accounts for 40 percent of the global glyphosate weed killer market. Fashion pro-environmentalists were skeptical that all this scientific innovation was good for Earth. In 1994, Patagonia founder Yvon Chouinard researched the impact of Roundup, concluded that it was toxic, and pledged to source only organic cotton by 1996. Twenty years later, Chouinard's concern was confirmed: in 2015, the International Agency for Research on Cancer, an arm of the World Health Organization, classified Roundup and other glyphosate-based herbicides as "probably carcinogenic to humans." In 2018, Bayer bought Monsanto and announced it would retire the century-old name, which, among activists, had become synonymous with corporate evil.

To combat cotton's myriad of scourges, farmers apply Bayer's Aldicarb, a carbonate insecticide. Aldicarb is one of the most widely used pesticides. It is also demonstrably poisonous to both humans and wildlife. Exposure to it can cause blurred vision, headache, nausea, tearing, sweating, and tremors; in high doses, it can be lethal. Burglars in South Africa have been known to use Aldicarb to poison dogs.

Worryingly, sixteen American states have reportedly found traces of Aldicarb in their water tables. Under President Obama, the Environmental Protection Agency drew up a plan to phase out Aldicarb in 2018. But in 2017, after the Trump administration took over the federal agency, the web page outlining the phaseout had not been updated.

DESPITE ALL THE FASHION ills that denim represents, its popularity carries on unabated. This cycle of fashion consumption continues for many reasons, not least because of "the pressure of capital markets, private and public," David Weil, dean of the Heller School for Social Policy and Management at Brandeis University, told me. The supply chain—from raw materials to labor—has been corrupted. "What's viewed as acceptable behavior has eroded," he said.

To change it, he believes state and federal agencies must bring "the tops of these companies to the table and get them to have a greater incentive to set pricing structures that behave differently."

Ultimately, he said, brands will have to accept the idea of smaller profits.

It also means "consumers will have to pay somewhat more," he said. "If consumers want $11 garments that they are going to feel good about, wake up."

In short, the industry needs a reckoning. Today, fashion executives "dictate everything they want in their supply chain to an incredible degree—a specificity not only of the product, but product delivery, barcode, shipping containers," Weil said. "They will send back an order when the dyes aren't right—they monitor *that* precisely.

"But," he continued, "somehow it's unreasonable to make sure that there is adherence to fire emergency escapes rules, and you aren't operating in a building like Rana Plaza. Either you start attacking that piece of this problem, through a combination of consumer and NGO pressure, and cooperation of governments."

Or, he said, you come up with "a different production model, entirely."

# part two

# FOUR

# Field to Form

IN THE NORTHWEST CORNER of Alabama, across the Tennessee River from R&B recording mecca Muscle Shoals, is Florence, a town of 39,000. Before NAFTA, Florence was the Cotton T-shirt Capital of the World.

"They used cotton that was grown around here," fashion designer Natalie Chanin told me, over heirloom BLTs and iced tea at The Factory Café, her farm-to-table restaurant located in Bldg. 14, one of twenty immense one-story former factories in an industrial park on the edge of town.

English-fair, with a plane of ash-white hair, happy hazel eyes crowned by crow-black brows, and a voice like Tupelo honey, Chanin is "eighth or ninth generation" southern and a Florence native.

She remembers clearly when her hometown was a robust apparel manufacturing center—where we were lunching once had been the home of Tee Jays Manufacturing Co., the third-largest employer in the Shoals area. In the early 1990s, pre-NAFTA, Tee Jays had an annual payroll of $50 million.

"There was actually a knitting machine in this building, and the dye house was back behind us," Chanin said. "This"—sweeping her hand around the wide-open space where we were sitting—"was a sewing floor. Just rows and rows of machinery. Hundreds and hundreds of hemmers. Ralph [Lauren], Tommy [Hilfiger], and Walt Disney all produced here."

After the passage of NAFTA, US T-shirt production moved offshore. Local manufacturers like Tee Jays ceased operations. "NAFTA destroyed the company," its former owner, Terry Wylie, told me. Florence, like much of the textile-driven South, plunged into financial and social crisis. "In 1993, five thousand worked in this two-block radius," Chanin said. "And that didn't include all the service industries—restaurants, daycare centers, gas stations. There used to be twenty dye houses in this town. When manufacturing collapsed here, everything collapsed."

Now, twenty-five years on, Chanin and her friend, Louisiana-born fashion designer Billy Reid, are helping Florence live up to its sobriquet of "Renaissance City."

At The Factory, Chanin and her team of thirty run Alabama Chanin, a women's wear brand specializing in flowing organic-cotton dresses and smart tailoring, all produced in the region. On Court Street, the town's main thoroughfare, Reid has his headquarters and a shop—one of twelve nationwide when I visited in 2018. His signature look is what a *New York Times* fashion writer described as "whiskey-soaked style": crinkly seersucker blazers, crisp linen pants, frayed work shirts, cotton chemise dresses, and high-quality selvedge jeans abound.

To staff their companies, Chanin and Reid have recruited a tribe of young urbanites—in addition to Chanin's thirty, Reid has seventy employees. That influx of creative sorts has spawned a slew of happening new businesses—gastropubs, boutique hotels, a microbrewery, and Single Lock Records, cofounded by local Grammy-winning musician John Paul White. Each August, Reid throws Shindig, a three-day celebration of southern food, music, fashion, and culture, open to the public, drawing visitors from across the

South. Chanin kicked off the edition I attended—Shindig No. 8, in August 2016—with a rollicking benefit dinner for a couple hundred at The Factory Café to support the Southern Foodways Alliance, a regional institute for the study of southern food culture.

What Chanin and Reid are doing is known as slow fashion: a growing movement of makers, designers, merchants, and manufacturers worldwide who, in response to fast fashion and globalization, have significantly dialed back their pace and financial ambition, freeing themselves to focus more on creating items with inherent value, curating the customer experience, and reducing environmental impact. This quiet revolution is also driven by their desire to improve the quality of life for their families and their employees.

Slow fashion champions localization and regionalism rather than mass-ification. It honors craftsmanship and respects tradition while embracing modern technology to make production cleaner and more efficient. It's about treating workers well, Chanin said, and "buying from the person down the street whose face you know and love."

She supports like-minded fashion folk as best she can—she has sourced cotton from Sally Fox, has collaborated with the young Nashville-based designer Elizabeth Pape of the direct-to-consumer Elizabeth Suzann label, and is good friends with and a sounding board for New York–produced women's wear brand Maria Cornejo. When aspiring designers reach out to Chanin to ask how she makes it work, she always tries to respond. "It's so important to give a hand to the next generation," she said. And every day at two p.m., she opens The Factory for public tours. "We try to be as transparent as possible."

She also believes education is key. At The Factory, she has opened the School of Making, an outreach program to teach sewing, and she publishes books on needlework. "Students who come to us from design school, all they know is drawing pictures, sending them off, and getting finished garments back," Chanin said. "There's a real lack of understanding of how clothes are made—a lot of critical knowledge that's been lost."

In 2001, Chanin produced a short documentary called *Stitch*, about the art of southern quilting, to show with her first collection. Since 2016, she has been working with the Center for the Study of Southern Culture at the University of Mississippi to record an oral history of sewing in the South. And in April 2019, she introduced some of the findings during her inaugural Project Threadway's Symposium, an annual celebration of "manufacturing, music, and community" with a focus on material culture, textile history, cotton, and women in the workforce. She sees her education initiatives as a way to "preserve" needlecraft—"a skill," she said, "that is dying out in this country." For Chanin, it's imperative: we should "be able to make our clothes." If we lose the knowledge of handcrafts like sewing, she asks, "What happens to the culture?"

CHANIN GREW UP around cotton fields. Her maternal grandfather worked for the Tennessee Valley Authority, her paternal grandfather was a carpenter, and both were farmers. "My mother always throws out that she picked cotton to buy her school clothes," she said with a laugh. Chanin's maternal grandmother and great-grandmother worked at the Sweetwater Mill in Florence, producing military-issue underwear. Her mother was a middle-school math teacher, and her father was a carpenter and a commercial contractor.

They all taught Chanin the importance of self-sustainability. Her grandmothers sewed at home—one "made everybody's underwear, nightgowns, everything," she said—and they showed her how, too. As a child, Chanin spent hours in her grandmother's attic, playing dress-up with vintage dresses, capes, and shawls. "That's how I fell in love with clothes," she told me. A young mother—she had her son, Zach, at twenty—she studied fashion and textile design at North Carolina State University. Upon graduating in 1987, she migrated to New York to work as a design assistant for a junior sportswear brand on Seventh Avenue. What she encountered there made her reevaluate her career dreams.

"I spent a lot of time overseas, and I saw a lot of things that I don't think are right—things that you don't want human beings to do," she said. She heard horror stories, as well. One friend, who worked for Gap, "told me she visited a dye house in India and the dye was just pouring directly into a river," Chanin remembered. "And ten feet down the river, kids were getting water and drinking it. They were drinking blue dye. The river was blue. I thought: 'If that's how I have to make fashion, then I don't want to make fashion.'"

In 1990, she moved to Vienna, Austria, and became a stylist for MTV. She returned to New York in 2000, for what was supposed to be a sabbatical, and took up lodging at the Hotel Chelsea. She spent her days at Goodwill, buying T-shirts. She'd cut them up, collage them back together, and decorate them with unusual embroideries that had exposed knots and dangling threads. An antique corset she picked up at the Twenty-Sixth Street flea market served as inspiration. "I couldn't figure out what was the inside and what was the outside, and it had been cut away and added to," she recollected. "That's how I felt of my life at that point—inside out and upside down and sideways—so I turned the T-shirts inside out, revealing the work on the underside. And that defined our style."

To make a proper collection, Chanin needed help. She called on workshops in the Garment District. "I was trying to get them to do the fancy embroideries, and nobody could understand what I was talking about," she said. "Then I realized it looked like a quilting stitch, and if I wanted to get it made as I wanted, I needed to come home to Alabama, where people still quilted."

She looked to rent a home in the countryside where she grew up, she said, "because I thought that is where I would find the quilters." Alabama has a long history of quilting, carried on by such associations as the Gee's Bend Collective and, until 2012, the Freedom Quilting Bee.

Chanin's aunt had recently purchased a redbrick house that Chanin's grandfather had built for his best friend in 1949, right next door to his own home. Chanin moved in, set up a few sewing machines and a desk, and found quilters to do her embroideries. Project Alabama was born.

When enough clothes were ready to photograph and sell, Chanin put together a small catalog, known in the biz as a "look book," and sent it to retailers. One landed on the desk of Julie Gilhart, then the high-profile fashion director for Barneys. Gilhart's forte was discovering and promoting young fashion talent. Intrigued by the look book—"It was artistic and amazing," Gilhart told me—she called on Chanin at the Chelsea. She liked what she saw. "Natalie's T-shirts had a lot of style to them, and her collection had purpose to it—she was employing a lot of women in Alabama and encouraged crafted culture," Gilhart recalled. "I loved that, and we ran with it." Several other retailers followed, including Ron Herman in Los Angeles and Browns in London. The T-shirts retailed for up to $400 apiece. "The designer customer was Natalie's patron," Gilhart said. "They sold very well."

From there, Chanin developed a more complete collection, with 1930s- and 1940s-style dresses and suits like Blanche DuBois wore in *A Streetcar Named Desire*. All were made in organic cotton or recycled materials, by seamstresses in Florence. Chanin jetted between New York and Florence, and twice a year she would take the collection to Paris Fashion Week and present it to international retailers in a Left Bank hotel room. Back then, 80 percent of her sales were overseas, and all were wholesale.

I first met Chanin about this time, in one of the Craig Ellwood–designed bungalows at the Chateau Marmont in West Hollywood; she was in town to drum up business during L.A. Fashion Week. I was invited over by a mutual friend, and I brought along my then four-year-old daughter. Chanin sat on the mid-century modern sofa, with cotton jersey in her lap, a threaded needle in her right hand, and the fingers of her left running down the strands, as if she were massaging them. "This is called 'loving the thread,'" a southern practice to prepare the thread for sewing, she explained to my daughter. During the spinning process, the fibers are twisted until they are taut. "When you sew and your thread tangles, it's because it has too much tension," she said. "One of the traditions I was taught is 'loving your thread': as you pull it through your fingers, the oils in your skin coat the strands, and you release the tension. Then it doesn't tangle as much." She demonstrated this to my daughter, and the two of them sat there, calmly and methodically drawing thread through their fingertips.

Chanin's sales went well enough, with wholesale accounts at about $2 million. "But as we all know, building a fashion business is hard," she told me. "And building a fashion business based on artisan handwork in the US is extremely difficult. Finally, in 2006, there came a point where ideas diverged and my business partner and I went our separate ways." Around the same time, she had her second child, a daughter.

She decided to start over, rechristening her brand Alabama Chanin.

"Same people. New name," she said.

And no more New York anything.

Chanin was ready to fully embrace what her friend John Paul White calls "the nurturing benefits of a small town."

She wasn't alone. Thanks to the internet, telecommuting had become mainstream. And with the development of inexpensive, and simpler, point-of-sales software—and smartphone and tablet apps—small businesses could afford to set up retail online. Until those digital advances came around, a hyperlocal movement seemed impossible. Places like Florence were too far away—too disconnected—from the New York-London-Paris-Milan design and retail network. But the technology that made globalization easier for giants like Zara and H&M to produce and sell so much also made it possible for small towns to become fashion hot spots. The status quo was ruptured, the power decentralized.

Chanin contacted Tee Jays's former owner, Terry Wylie, who still held Bldg. 14, and negotiated to take over a 20,000-square-foot portion of the 160,000-square-foot building. She packed up operations at the redbrick house and moved in. Her new space was like a time capsule from the early 1990s. "There was still a pay phone on the wall," she remembered.

She has since doubled that original space to 40,000 square feet. Her studio, walled off by plywood and corrugated metal partitions, is outfitted with worktables, a dozen sewing machines, and a library with hundreds of books on American craftsmanship, sewing, and southern cooking. She has an extensive archive of the jerseys and embroideries she has developed over the years and reaches into it regularly for inspiration. "I've often thought that one day we'd travel to Paris with these," she said, as she handed me a few embroidery swatches, "and say, 'These are the things we can do. Send your work to us!'"

She buys her fabrics from Signet Mills in Spartanburg, South Carolina—her preferred is jersey made of organic Texas cotton—and she works with a local artisan dyer for her indigo pieces. Her dedication to organic materials contributed to her winning, in 2013, the CFDA/Lexus Eco-Fashion Challenge, an award that heralds as well as encourages the development of sustainable fashion.

Chanin comes up with the designs, and her assistants execute them. When I visited, her sample sewer was a sixty-eight-year-old, pre-NAFTA garment employee named Sue Hanback. "She essentially came out of retirement to help us," Chanin said. "We could not have done it without Sue." Hanback has since retired again but still lends a hand when needed.

Rather than print on her fabrics like Mary Katrantzou does, Chanin uses stencils, "a universal form of pattern transfer," first developed by the Chinese in AD 100, she said. "We have about seven hundred right now, some geometric, some floral. Made from a variety of materials from pasteboard to very sturdy Mylar that can be used hundreds of times." The patterns are spray painted on the cloth with an airbrush gun.

Once the fabric is cut and stenciled, and the embroidery materials and notions are chosen, the team bundles the whole lot up in what Chanin calls a "kit." When a customer places an order online, Chanin's freelance seamstresses—about two dozen in total—bid for the job. They are all independent contractors, free to decide when, where, and for whom to work, and they build extra costs, such as supplies, utilities, healthcare, and other benefits, into their bid. She awards the project based "on timeliness or quality of work," she explained. The contracted sewer drives to Bldg. 14, picks up the kit, stitches it in a day or two, numbers and signs the finished garment, and brings it back to the Alabama Chanin HQ to be packaged and sent off to the customer. A hand-sewn double-layer organic cotton jersey dress retails for about $800; Chanin's hand-sewn organic cotton coats are almost $4,000. Of that, her sewers earn a minimum of 25 to 50 percent of the full retail price, depending on the complexity of the project. "I was sitting at a dinner last night at friends of mine, and they said, 'Your things are so expensive!' And I'm like, 'Fuck, yeah, they are. Because I pay my people right.' I'm not driving a Mercedes. I've got a Prius. I have a very modest life."

Chanin does what she can to avoid labor issues that have long dogged the southern apparel industry. "There are a lot of people who ask, 'How do you know it's not child labor?'" she told me. "And I say, 'Well, we've known Miss Betty now for sixteen years. She's eighty-six years old. She only does one particular kind of work. She doesn't have any kids at her house crocheting snap covers.' And it's why we have a rule that our sewers have to all live within an hour and a half of us; you have to pick up and drop off your work yourself. If somebody's coming in and taking fifty kits a week, it's a pretty good sign that they're not doing it themselves. It's just a more personal relationship. And it's all women."

"It's *all* women?" I asked.

"It's *all* women," she said.

In 2013, Chanin added machine-made to the mix; those clothes are all produced at Bldg. 14 and run between $59 and $1,000—"depending on intricacies," she explained. It was a smaller slice of her business than she would have liked. "Our production capacity is limited by finding sewers who can run the machines to a quality standard," she said. To train more machinists, she has partnered with NEST, a New York–based nonprofit that supports artisan fashion communities throughout the world. Chanin also sells DIY kits—which range from $150 for a T-shirt to $550 for a wrap dress—for customers who want to sew the clothes themselves.

About the same time, she opened the café and the shop. The café is made up of long rows of white-painted wood tables with mismatched farm chairs and serves lunch six days a week. On the day I visited, the chalkboard behind the bar read: "Welcome friends." The shop, an open space that flows into the café, carries covetable southern artisanal homewares, as well as a selection of Chanin clothes. A snippet of Chanin's business is now wholesale to places she likes or that are run by friends, such as the Blackberry Farm hotel in the Tennessee Smoky Mountains and Smilow Mathiesen gallery in Santa Fe. The majority of her sales, however, are online and made-to-order, usually with a three-to-six-week delivery time. When I visited her in 2016, 60 percent of her business was e-commerce; two years later, it was 80 percent.

Chanin's "lean method of manufacturing," as she calls it, produces roughly 120 garments a day—a raindrop compared to the 35 million the Tee Jays factory put out a year in its pre-NAFTA heyday. In 2013, she drew up a ten-year plan, with the goal of reaching $10 million in sales by 2023. In 2018, her turnover was $3 million, and she said, "We are spot-on for our annual goals."

Implementing a more ethical business model hasn't been "the most lucrative" way to run her company, she conceded. And there were plenty of Cassandras along the way who told her that having an unbroken domestic supply chain was impossible. "But we've stuck to our standards, even when it wasn't the easiest thing to do," she said. "And we've made it."

Sure, there are times when, by doing everything in northwest Alabama, "we miss the deeper connection to the industry and the heartbeat of what's happening in design in America," she admitted. But the advantages of being "hyperlocal," as she calls it, outweigh those wistful moments of longing. When the "difficult times come around"—and they do—"our overhead and expenses are so low, it's not as frightening," she said. "I'm 100 percent self-owned—no partners. We don't owe the bank. We don't borrow money to produce the collection. We invest in young people and train them well. We have a deep commitment to our community. I have been able to raise my children and live a creative life that makes me happy and do good and important works. I like where I've landed and what we have created. And I'm proud of having been active in bringing something back to my hometown and contributing to its future."

LIKE CHANIN, Billy Reid had to stumble to realize that small-town slow fashion was the way to go.

Born in 1964 and raised in Amite City, a Louisiana bayou outpost an hour and a half northwest of New Orleans, Reid is a second-generation fashion peddler; his mother, T.J. Reid, had a dress boutique in his grandmother's former home. "I always describe it as '*Steel Magnolias* in a shop,'" Reid told me during my visit to Florence. "My mother didn't care if customers shopped; they might just come to talk and gossip."

Reid had grander ambitions—and for a while, he realized them. In 1998, he founded William Reid, a Dallas-based label that concentrated on wholesale. Within two seasons, he had thirty-five accounts, including Saks Fifth Avenue. In 2000, he moved the company to New York, setting up in a warehouse space on Twenty-Eighth Street, between Tenth and Eleventh Avenues in Manhattan. He dressed celebrities like Matthew McConaughey and Gwyneth Paltrow and was touted by *Vogue*. In June 2001, he won his first of four Council of Fashion Designers of America (CFDA) citations: the Perry Ellis Award for Emerging Talent.

His Spring-Summer 2002 fashion show was staged in his headquarters on September 10, 2001. "Great show," he recalled. But no one had a chance to read the reviews or buy the clothes. The terrorist attack the next morning also clobbered the American economy, including Reid's fledgling business. Retailers canceled appointments and orders, and his new financial backer withdrew a $10 million pledge. Six months later, he was forced to fold. "We lost everything," he told me.

He and his wife, Jeanne, retreated to her hometown of Florence, two small children and two big dogs in tow, and moved into her parents' house. He tried to relaunch his business in New York, but that went nowhere. His friends and associates Katy and K.P. McNeill had another idea: create a life-style brand that embraced Reid's southern roots. They wrote up a business plan and presented it to him. He liked it, and together in 2004 they launched Billy Reid, based in Florence, with K.P. as chief executive, Katy as chief merchandising officer, and Billy as creative director. "Being in Florence differentiated us right off the bat," K.P. told me in 2016, as we walked through the company's offices above Reid's shop, in a handsome early-twentieth-century building on the town's main street. "If Billy were just another de-signer in New York, it'd be so much tougher. I don't think we'd be able to operate profitably if we were there."

The greatest difference between the defunct William Reid and the ever-growing Billy Reid is that the 2.0 version is foremost direct-to-consumer, either in his self-owned boutiques or online. Wholesale was limited from the start and has diminished with each year. When I met Reid in 2016, the breakdown was: 60 percent direct-to-consumer, of which 15 percent was e-commerce; and 40 percent wholesale, half to department stores and half to specialty boutiques such as Oak Hall in Memphis and Shaia's in Birming-ham. "Old-school stores, who know what they're doing," K.P. said.

The shift away from department store chains is not exclusive to Billy Reid. In recent years, retail institutions such as Macy's, Lord & Taylor, and Neiman Marcus in the United States, and John Lewis and House of Fraser in the UK, have reported tumbling sales and profits and closed stores by the score—leaving suburban malls sans anchors. "Department stores will be gone in the next three to five years," McNeill told me—echoing a lament I've heard repeatedly from many fashion folks. "It's a fundamental change in how business is done. Direct-to-consumer is the future."

As with all things in business, efficiency and profit margins have driven this shift. Back when Reid sold exclusively wholesale, he had to design and make 250 to 300 looks each season to present to retailers during Fashion Week; of that, buyers would select about a third—generally the most bland and easy to sell, like black pants and white shirts—and Reid banked about

30 to 40 percent of the retail price. The remaining samples were scrapped. With his vertically run company—meaning he has a hand in everything, from design to retail—Reid creates as he pleases, offers more daring items, avoids redundancy, controls distribution, and pockets 60 to 70 percent of what the customer pays, which he plows back into the company to produce better-quality clothes. Three years after its debut, Billy Reid was in the black. By 2017, the brand counted $25 million a year in sales.

"I think the people who are going to be successful in the US today are those who are completely vertical—who offer an entire culture and vision," McNeill said. "You aren't just sewing a shirt for someone who is beating you up on price. That's not it. As a matter of fact, you *can* pay a little more. Big companies have this mentality of third-party cut-and-sew outsourcing. But there's actually a way to make more money long term with this new model— this new approach.

"The holy grail is: here's the order, and in twenty-four hours, the garment is made and shipped," he continued.

"Like fast fashion," I said.

"In a quality way," he allowed. "There are certain things that can't be done. You're not going to get a jean through a wash process and embellish it all in a day. But if you have the fabric ready, the garment's made and goes out the door. There's no wholesale. No real estate cost. No inventory cost, and no inventory risk."

"We broke down . . . barriers," Reid said. "You can do it from anywhere if you do it right and do it real."

IN 2011, K.P. McNeill was driving past some local cotton fields during harvest when he had an epiphany: why not "go from seed to finished product in the same community?" The ultimate in vertical integration.

And why not do it in Florence?

After all, the region had a deep-seated textile tradition, cost would be reasonable, and there was still the local knowledge of how to go about it, the machinery to turn the earth, and thanks to Reid and Chanin, a swarm of young creative talent gung-ho about all things sustainable and Americana. McNeill told Reid and Chanin his idea, and they loved it.

Back before NAFTA, Chanin said, the local textile and apparel businesses "were growing the cotton; they were ginning the cotton; they were processing it." They went "straight from field to form."

Reid and Chanin wanted to return to that business model, but with a modern twist: it would be organic. They reached out to the Sustainable Cotton Project in Winters, California, for growing tips and the Texas Organic Cotton Marketing Cooperative for seed. Today, 99 percent of all cotton is genetically modified.

"We found out that there is a dearth of [organic] cottonseeds on the planet," Chanin said. "We went on this search for months. It was really scary."

Eventually, they collected enough to plant their first crop. There were doubters. As Chanin noted, "All the farmers were like, 'This is not going to happen. You can't grow cotton here.'"

"So many people were betting against us and saying, 'You can't grow cotton unless you use pesticides. The bugs will eat it. It will be gone. Good luck, *ha ha*,'" said Lisa Lentz, who, with her husband, Jimmy Lentz, owned the project's farmland. "This little cottonfield was planted just like our grandpas would have planted."

"We had a drought. We didn't water. We didn't do *anything*," Chanin said. "Weeds started coming up, and we weeded by hand, but in some sections the weeds just kind of took over. And the cotton still grew and thrived."

For the harvest, in the spring of 2012, "people drove from all over," she told me. "Some folks flew in from San Francisco. And we had a cotton-picking party. Six acres. Six hundred pounds. Three hundred people. They were singing together, laughing together."

The cotton was bagged and sent to a local gin, Scruggs & Vaden, to remove the seeds. Then it was shipped to Hill Spinning Mill, a fifty-year-old mill in North Carolina, to be spun into thread. The mill's machines were sanitized before the organic cotton was processed, so the chemicals from traditional cotton wouldn't contaminate it. The mill owner said he had never seen such clean cotton—the result of handpicking.

Some of the thread was sent to Gina Locklear of the Little River Sock Mill in Fort Payne, Alabama, to make socks. Locklear—known as the Sock Queen of Alabama—is a second-generation mill owner. Her parents, Terry and Regina, opened the plant—named Emi-G, for Gina and her sister, Emily—in 1991 and made white sport socks for Russell Athletic. Back then, Fort Payne, population 14,000, was known as "the Sock Capital of the World," and its more than 150 factories produced one out of every eight pairs worldwide. Then the Central America Free Trade Agreement (CAFTA) went into effect and Fort Payne lost its business to Honduras. The Locklears held tight, even though they had next to no orders. They knew if they closed shop, they'd never reopen. "We'd just come here and sit," Terry Locklear said.

In 2008, at the age of twenty-eight, Gina stepped up. An avid environmentalist, she wanted to combine her passions for sustainability and socks. At the family mill, she produced a line of organic cotton socks she called Zkano, pronounced "za-ka-no," an Alabama Native American word that "loosely translates as a state of being good," she told me. The colors are bold, with jazzy graphic designs. In 2013, she introduced a second line, Little River Sock Mill, an Americana-inspired collection in softer tones, tamer patterns, and vintage floral designs. All sold well; the factory was vital and bustling again.

Locklear had already collaborated with Chanin and Reid on a few projects. So when Chanin reached out for this one, "Oh, I was excited!" Locklear told me. "I have admired Natalie and Billy for a long time—before I started making socks—because of the positive light they were shining on our communities and our state. And now they were growing cotton in our state, too.

That was wonderful." The couple of hundred pairs of socks Locklear produced were natural—no patterns, no dyeing—in basic shapes. "Our technicians were really impressed—they said it ran really well," she recalled. "I remember them saying: 'Great cotton.'"

The rest of the Chanin-Reid cotton was woven by Green Textile (now Signet Mills) in Spartanburg, South Carolina, into cloth—about seven hundred yards' worth—and sent back to Florence, where it was fashioned into clothing. "It took us about a year to take the cotton through the whole cycle," Chanin said. "We proved it could be done."

She gave me one of the V-neck T-shirts. Made of a soft, dense vanilla jersey that is solidly seamed, it is one of the best-made, most comfortable shirts I have ever owned.

In 2018, K.P. and Katy McNeill left Billy Reid and Florence for a new adventure, two hours by car north, in Nashville: they became the new owners of Imogene + Willie—pronounced "Eye-muh-gene and Willie"—a local casual wear brand championed by fashion cognoscenti for its retro, butt-cupping jeans, sewed, at least in the early days, in an atelier in the shop.

Imogene + Willie is a cautionary tale in the slow fashion movement: it grew fast—*really* fast. Its owners moved operations to L.A., forsook their original mission of hyperlocal for something bigger and more industrial, and lost their way. The brand was on the verge of extinction, less than a decade after its founding, when the McNeills brought it back to its original home, a restored 1950s filling station in the happening 12South neighborhood of Nashville. Adhering to the slow-fashion, direct-to-consumer model, they axed wholesale, only offering product in their one store or online. By the end of their first year—2018—sales were hovering around $3 million and on the edge of profitability, with a goal of $10 million in five years. Their philosophy: "Take small steps, not grow just to grow," K.P. said. While much of the line is still made in Los Angeles, eventually they'd like to produce in greater

Nashville. Keeping manufacturing close, he said, will allow them "to make sure everything we do we can be proud of."

Nashville has long been America's Music City. But it is also the third-largest fashion center in the United States, after Los Angeles and New York. Much of the output is government contracts—military uniforms, mostly, since federal law forbids the offshoring of such work. And, of course, there has always been a significant costume business for the local entertainment industry—think of the wildly embellished getups made famous by Elvis, Dolly Parton, and now Jack White.

In the last decade, however, straight-up fashion has become a prominent third segment. As with Florence, Nashville's primary draw is economic; like in Florence, the cost of living and working is considerably less than in New York or Los Angeles. The oft-cited statistic is that more than one hundred people are moving to Nashville a day—double the national average—making it one of the fastest-growing cities in the United States. But there are other tempting lures for apparel companies: an easily accessible international airport, no state income tax, an abundance of lakes and rivers (water is a necessity in apparel production), and the city's culture of nurturing and celebrating creativity.

In 2017, the Nashville Fashion Alliance (NFA) reported that fashion accounted for 16,200 jobs and $5.9 billion in revenue in the region and predicted that by 2025, those figures would jump to 25,000 and $9.5 billion, respectively. More than half the companies—one hundred-plus—had been established in the last five years. In addition, each spring, the not-for-profit Nashville Fashion Week hosts a slate of runway shows, panel discussions, and happening fetes. In short order, Nashville has become a viable force in American apparel.

Like Imogene + Willie, most Nashville fashion brands specialize in the town's everyday look: chambray shirts, good cotton T-shirts, selvedge jeans. (Nashville is *not* a suit city.) But there's a host of cutting-edge designers who have a broad following. For example, former Billy Reid menswear head

Savannah Yarborough started her own bespoke leather jacket brand, Savas, in 2014. And Ceri Hoover is known for her artisanal handbags and shoes. And there is Elizabeth Pape, founder and designer of the much-lauded fashion brand, Elizabeth Suzann.

Pape is Nashville fashion's rising star. A self-taught sewer, she started Elizabeth Suzann—her first and middle names—in her spare bedroom in 2013. When I met her in 2016—at the tender age of twenty-six—the company had grown to $3 million a year in sales, with eighteen employees in a ten-thousand-square-foot space in an industrial park on the edge of town. Her clothes are minimalist chic: soft jackets, slouchy pants, and simple shifts in neutral-tone wool, silk, linen, and cotton—all sustainably and, if possible, domestically sourced. In 2018, her prices were running from $125 for a sleeveless linen crop top to $265 for a canvas coat. And everything is made to order, with delivery in two to three weeks. "We don't have an inventory," Pape told me in her showroom. "No leftovers."

Like at Chanin, each sewer makes the garment in its entirety, down to stitching in the label. For the mix, Pape has adopted the fast-fashion strategy of having a signature collection of seasonless staples, punctuated by frequent drops of new designs. That way, she said, habitués "come back to us." She sells best in New York and Los Angeles but is content not to be based in either. "It's nice to feel a little bit separate and a little bit independent," she said. "It helps me to not feel overwhelmed and intimidated and just part of this big machine."

The only obstacle Pape has faced in Nashville is the lack of experienced hands. NAFTA wiped out the garment production labor force twenty-five years ago; folks who lost those jobs have gone on to other careers, are retired, or have passed away. Home ec classes disappeared about the same time. These days, Pape said, "young people aren't really interested in sewing." Everyone on her staff she trained herself. Van Tucker, former Nashville Fashion Alliance head, described the needlecraft shortage as "one of the biggest challenges" in Nashville's burgeoning apparel industry. To make up for the deficit, the NFA and Catholic Charities of Tennessee have established an academy to teach refugees how to sew. Like a hundred years ago in New York, sewing jobs are a starting point for immigrants pursuing the American dream.

At Omega Apparel in Nashville—a company that had survived the NAFTA exodus by securing military contracts and, when I visited in 2016, was moving into streetwear such as T-shirts and hoodies—the entire production team consisted of immigrants and refugees placed by the Catholic Charities program. Most came from political hot spots like Myanmar, Iran, Syria, and the Sudan. Several of the women were Muslim, wearing traditional headscarves. "We communicate through Google Translate," one of the managers explained to me. "But they are learning English." A bell rang, signaling lunchtime. The sewers sat down at a long table next to their work area and opened lunchboxes, and the room filled with the aromas of international cuisines.

In early 2020, apparel entrepreneur David Perry aims to open a new knitwear factory in Nashville. A British transplant in his early fifties, Perry has produced premium knits and activewear in Los Angeles since 2007. He relocated to Nashville to follow his wife, Leigh, a Louisville, Kentucky, native and a singer in the Americana duo the Watson Twins. Nashville has since won him over both on personal and business levels; at the time he moved, "the *living* wage here [was] less than *minimum* wage in California."

Starting *ex novo* gave Perry the freedom to design a manufacturing center of his dreams. Solar-powered and staffed with workers paid several dollars an hour above the federal minimum wage, he believes it will be "Tennessee's first fully transparent, fully sustainable factory." At first, he will bring in some of his workers from L.A. "They are excellent speed sewers, and if you're going to start a race team, you need the fastest driver," he said. He plans to help them settle into the community well, then he'll hire local labor, which his ace L.A. team will train.

The factory also will house a design center, a fabric showroom, sourcing consultants—"everything except dyeing," he said. "If you are a New York brand and you want to produce in the US, you can come here and we will help you, from start to finish. Production will be on your doorstep, instead of two thousand miles away in L.A., or overseas." Both Imogene + Willie and Billy Reid have expressed interest in sourcing there.

"The L.A. garment manufacturing industry was built on a broken business model—primarily on workers who weren't paid a fair wage, or were illegal," Perry said. In Nashville, "we don't have a broken industry to fix. We have the opportunity to build an industry the right way. I will tell customers: 'This is my price, this is the right way to do it, and we will give you the ability to proudly support American jobs and ethically practice.' I want clients to march in and say, 'This is fantastic. This is how manufacturing *should* be.'"

# FIVE

# Rightshoring

THE FOLKS I visited in Florence and Nashville showed me that slow fashion on a moderate scale—the $5- to $10-million-in-revenue-a-year size—was possible, profitable, sensible, and enviable. But could their hyperlocal, thoughtful approach to business be scaled up from cut-and-sew workshops to the level of factories, where there are hundreds, or thousands, of workers and assembly-line production? Could this be a way to reignite domestic manufacturing in developed economies? That's what I was about to find out in Cottonopolis, of all places.

On a mouse-gray November morning in 2016, I ventured east of Manchester, England, to Stalybridge, a mill town Friedrich Engels described, in 1845, as "repulsive," consumed by "shocking filth." Tameside, the borough that contains Stalybridge, is full of disused Victorian-era mills that have been converted into apartments, offices, supermarkets, even gyms. Tameside has cleaned up. It has become a middle-class suburb.

It has also become a center for cotton milling again: nearly

seventy years after it had ceased production, Tower Mill, a redbrick mono-
lith with an imposing smokestack, was spinning *Gossypium* for a new company called English Fine Cottons.

Commercial director Tracy Hawkins welcomed me. A hearty blonde in
her early fifties with a long history in the British apparel industry, she confessed she was extremely tired. In six months, "we built a modern mill from
scratch," she said. "But it's done, and it's working."

I happened to visit on English Fine Cottons's second full day of operation
and witnessed the first large-scale cotton spinning production in the United
Kingdom in more than three decades.

English Fine Cottons began as an afterthought. Manchester-born-and-
raised entrepreneurs Brendan McCormack and Steve Shaughnessy own the
historic Tame Valley Mill in Tameside, where they spin the technical yarn
for Kevlar. (While the UK apparel industry's textile business had long ago
died, industrial tech fabrics are still in demand, and easier to scale in developed economies.) Over the years, they also received requests to process cotton, but they always begged off, since it wasn't their business. Then, in 2014,
Tower Mill, which sits directly across the street from Tame Valley Mill, came
up for sale.

Designed by the celebrated Victorian architect Edward Potts, the four-
story mill was built in 1885, and at its peak, it counted forty-four thousand
spindles. Since going silent in 1955, Tower Mill has housed several businesses
and served as a set location for the early 1990s BBC television series *Making
Out*; in the early 2000s, there was talk of transforming it into luxury condo-
miniums. The plant went on the market just as McCormack and Shaughnessy were contemplating how to expand their company. They thought:
maybe we *should* get into cotton. But they didn't want to reshore the old
milling model with its Dickensian horrors. "We wanted to create a place
of excellence, producing well-crafted yarns," Hawkins told me. "Not a
museum."

With a private investment of £2.8 million ($3.65 million) from parent company Culimeta-Saveguard, a £2 million ($2.6 million) loan from the Greater Manchester Combined Authority's investment fund, and a £1 million ($1.3 million) grant from the Textile Growth Programme, the partners bought Tower Mill, restored it, and refitted it with the latest technology. Hawkins tackled the business strategy and concluded there were two keys to success: "flexibility," she said, "and quality."

Flexibility meant ensuring the mill could "spin yarns from the thickest counts to the finest—for weaving, knitwear, socks, anything that's needed," she explained. "Why would we want to set ourselves up as being the only cotton spinner in Britain, regenerating an entire industry, just to say, 'Oh, we're not interested unless you want to buy five hundred tons'?" As for quality, she wanted to guarantee "the artisanal way of approaching the business, provenance, heritage, and Britishness," she said. They would source only the finest cotton—at first, they considered organic but found its quality uneven and its quantity in short supply. Ultimately, they chose Supima, a sustainable US-farmed superfine long-staple, and the ultrasoft Sea Island strain from Barbados. "The classic English shirt used to be made of Sea Island cotton," Hawkins said. "It was what Ian Fleming said James Bond's shirts were made of."

The next challenge was to find a spinning expert to operate their sophisticated machinery. "How were we going to get somebody who could build and run a modern mill?" Hawkins wondered. They were not exactly marketable skills in Britain in the twenty-first century, but she heard that Paul Storah, a Yorkshireman who specialized in mounting and running modern cotton mills in South Africa, had recently moved back to the UK. She snapped him up as operations manager. "A little bit of stardust was sprinkled there," she told me, smiling.

The mill is a labyrinth of industrial chambers overrun with snaking conduits and whirring apparatuses. In the pretech days, owners relied on the Manchester region's damp climate to keep down the "fly," as airborne filaments are known. Today, the cotton travels through a circuit of enormous hydraulic tubes from one processor to the next. Most of the processors are self-contained: blowers that separate the fibers; blenders that mix the various grades of cotton; combers that "nip," or extract, the short filaments; or, my favorite, the "foreign-particle remover," a big glass box that looks like an oversized cinema popcorn-maker stuffed full of cotton. Lasers scan the nebulous mass for rogue seeds, leaves, or twigs, and when a piece of debris is detected, a needle-precise air jet blasts it out.

All is controlled by lab techs at computers in a sterile room. The air in the facility is changed out twenty-five to thirty times an hour. Modern cotton spinning, Hawkins told me, is "all about having clean air."

The cotton is sped through various processers until "it has a fantastic luster, it's light, and it's straight," she said. At each stop, it was purer, fluffier, prettier—exactly how you dream about cotton. Only at that point is it spun, at a dizzying 15,000 to 20,000 rpms, into a yarn. Fly is emitted—no way around that. But instead of relying on the damp air to tamp it down, or, as in the nineteenth century, children with little brooms to sweep between the spindles, the spinning machines are equipped with custom vacuum robots.

The ivory yarn is sent to Blackburn Yarn Dyers, one of Britain's last traditional dye houses, north of Manchester, to be tinted. Hawkins showed me spools of finished product in gray, navy, and putty white. It was floss-thin, smooth, and handsome. John Spencer (Textiles) Ltd., a sixth-generation weaver in nearby Burnley, Lancashire, that survived offshoring by becoming Britain's only certified weaver of organic cotton yarns, would turn it into socks for the UK mass retailer Marks & Spencer. By 2018, English Fine Cottons was producing—via a contracted weaver—its own namesake label cloth. Among its clients: Marks & Spencer (for men's shirting) and Aquascutum (for outerwear).

Despite all the automation, McCormack and Shaughnessy managed to create more than one hundred jobs at Tower Mill. English Fine Cottons produced 100 tons of yarn in its first year of business, and up to 450 tons in 2018.

When I last spoke to Hawkins, in the fall of 2018, she said demand was so high they had to hire more hands, and the mill was running 24/7. At that point, English Fine Cottons was the only large-scale cotton mill in Britain. It supplied weavers throughout the country; Burberry was buying its yarn, as was Peter Reed—"for the best bed linen you get in the UK," she said, proudly.

RESHORING—the act of bringing back the manufacturing that went offshore during the post-NAFTA globalization boom—has been gaining momentum in the last few years, especially for fashion. In the United States, textiles and apparel was the third-most-reshored sector in 2014, after electrical and transportation equipment, and the second-fastest-growing in 2016, employing 135,000 and putting out 10 percent of America's fashion— an impressive vault from 3 percent in 2013. In Great Britain, apparel production jobs jumped 9 percent from 2011 to 2016, to a total of 100,000, and another 20,000 were expected to be added by 2020.

Today's reshoring movement is *not* in the spirit of the "Make America Great Again" mantra—the Panglossian notion that if companies would re-shore what they offshored in 1990s and 2000s, all the laid-off employees could have their old jobs back. The fact is most of those workers "aged out" during the intervening two and a half decades. And perpetuating the same tired Arkwright business model doesn't take fashion forward.

No, what English Fine Cottons and its confreres are doing is "rightshor-ing": the reboot of domestic production—often in long-dormant factories— with state-of-the-art technology and transparency. Rightshoring looks "different than manufacturing looked in the 1980s," former Nashville Fash-ion Alliance head Van Tucker told me. "Innovation is going to snowball, and quickly. Social issues, especially sustainability, are important. And it's tech-driven. Very automated."

Paul Donovan, global chief economist at UBS Wealth Management, calls the trend "a reversal" of globalization as we have known it. "Robotics and digitization mean we can produce efficiently, locally . . . Trade wars today are fighting battles from the past." He predicts that global trade of goods, such as apparel, will "revert to something like the old 'imperial model' of importing raw materials and then processing it close to the consumer." Rightshoring doesn't mean mills must source everything nearby—cotton won't be grown in the UK for English Fine Cottons. But it does mean manufacturing near the ultimate consumer, like English Fine Cottons spinning yarn for Marks & Spencer socks sold throughout Great Britain, or designers sourcing fabrics close to home, like Natalie Chanin ordering jersey woven in the Carolinas.

Counterintuitively, technology may finally transform textile and apparel manufacturing to something more personal and ethical. Clothes don't have to be made by poorly paid, poorly treated workers using outmoded machinery. They can be produced in vertically integrated communities, on clean, quiet factory floors controlled by tech-trained assistants. And it can happen in communities that had long ago given up on manufacturing. Those assistants could be people you know. Or maybe the plant will be just up the road. Automation may not create thousands of manufacturing jobs, but the ones it does create—a hundred here, a hundred there—will be good, safe, and well-paying. Strange as it sounds, technology will bring humanity to the supply chain.

Rightshoring has so reinvigorated North Carolina's textile and apparel industry that, by 2017, it employed 42,000 in 700 factories, many of which are as high tech as English Fine Cottons's. In South Carolina, Parkdale Mills was one of America's biggest spinning houses for much of the twentieth century. In 1980, it would have taken 2,000 workers at the Gaffney-based plant to produce 2.5 million pounds of yarn a week. After the 2005 WTO-China deal, Parkdale moved operations to mainland China, and the town's mill went dormant. In 2010, Parkdale reopened in South Carolina with state-of-the-art machinery run by technicians from clean rooms overlooking the spinning floor. "We knew in order to survive we'd have to take technology as far as we could," Parkdale's CEO, Anderson Warlick, explained. Now 140 people are able to produce 2.5 million pounds of yarn a week. While that's not the 2,000 jobs of yesteryear, it's 140 more than zero. Success like this can be contagious—new businesses begetting more new businesses—until there's a veritable boom and long-dispirited towns are vital again.

The financial investment comes from everywhere—China included. In 2015, the Zhejiang-based Keer Group opened its first factory outside of its homeland, a $218 million, 165-acre "textile campus" in Lancaster County, South Carolina, that created more than five hundred jobs. Like Tower Mill, it is hypercomputerized.

Keer Group's primary motivation, of course, was economic: rising labor and energy costs in China had made milling a less profitable endeavor. But also—as McCormack and Shaughnessy learned—outfitting an empty factory with new technology is much easier than upgrading an operational one. There are no lost days of business, no layoffs, no junking of existing machinery. South Carolina had additional enticements, such as "proximity to cotton producers, and access to the port," Keer Group's chairman Zhu Shan Qing explained.

And, as McCormack and Shaughnessy discovered, there are "incentives" to rightshore—about $20 million worth in Keer's case, including infrastructure grants, revenue bonds, and tax credits. Today, there are dozens of high-tech Chinese-owned textile factories across the Carolinas—a fact that has some locals reeling in disbelief. As Lancaster County Economic Development Corporation's then-president Keith Tunnell confessed at the time: "I never thought the Chinese would be the ones bringing textile jobs back."


IN NEW YORK CITY, the task hasn't been simply to rightshore; it's been to shore up what bit of manufacturing remained. Some attempts have been more successful than others. Back in 1997, Brooklyn's borough president un-veiled a plan to establish a fashion "incubator" at Bush Terminal in Sunset Park; it languished ad infinitum. In 1998, the Garment Industry Develop-ment Corp., a management-labor organization, inaugurated the Fashion In-dustry Modernization Center; it eventually fizzled. Designer Nanette Lepore, whose company was based on West Thirty-Fifth Street, organized "Save the Garment Center" rallies and lobbied Washington lawmakers for support. Not a lot was forthcoming.

The most effective action, it turns out, has been the loyalty of New York–based talents like Maria Cornejo (pronounced "Cor-nay-ho"), of Zero + Maria Cornejo. Back in the late 1980s, the Chilean-born, British-educated designer worked for UK mass fashion retailer Jigsaw in Paris and saw firsthand what she calls "the false economy."

They'd send her to Hong Kong, business class, and put her up at the five-star Mandarin Oriental Hotel. Then they'd "nickel-and-dime" the custom-ers she was going to see. "They'd say, 'Well, we're going to save a dollar on a sweater, but we're going to ship it halfway across the world,'" she told me, in her Spanish-inflected English accent, as we sat in her book-stuffed office on Bleecker Street. "It didn't make sense to me. It just didn't."

Production was "a crazy system," in which the designers in Europe and the US would digitally send garment specs to a factory in Asia for the sam-ples to be made. After much back-and-forthing, via email and phone, the finished samples would be shipped to corporate headquarters to be reviewed and, more often than not, rejected. One major New York brand—a house-hold name—would order six hundred samples in China, then "cut it down to two hundred," she said. "Can you imagine the waste?"

"I just wanted to make everything in one place and, even if it was a T-shirt, have control from beginning to end—how it looks, how it was made, who made it," she explained. "I wanted to *know* those people." She wanted to rightshore.

In 1996, she moved with her husband, the photographer Mark Borthwick, to New York, took over an old garage on Mott Street, and in 1998, opened a shop with a sewing atelier in the back, to make and sell "interesting, easy clothes that you could afford to wear," she said. She called the brand Zero because she "wanted it to be just about the product, and not for people to have preconceived ideas of what it should look like or who was behind it," she said. On her first day of business, she made $2,500. "We thought it was a good sign," she later recalled. When she learned there was already a big German fashion company called Zero, she rechristened hers Zero + Maria Cornejo. (The plus is said as "plus.")

Her team was tight and international. Jiang Huang, an immigrant from Shanghai, sewed samples. (He later learned pattern cutting and took over that role.) Tonya, from Russia, handled knitwear. Lynn, from China, was in charge of silks. "She introduced me to Mr. Huang," Cornejo said. "And I cut everything." The labels read: MADE AT 225 MOTT STREET. And she sold the clothes in her boutique out front. "I remember one day a lady arguing with me about the price of my clothes, and I said, 'You see all those people working in the back? They live in New York. They get paid fair wages. We pay this rent. It's not like the clothes that have been sent to Timbuktu to be made by child labor.'"

Soon, like for Natalie Chanin, Barneys came calling, and before long, Cornejo had an impressive following, including Michelle Obama, Tilda Swinton, and Cindy Sherman. She kept growth in check, sticking only to women's wear and a few accessories, such as belts and shoes. "I've never been interested with having a pair of underpants with my name on them," she said. "It doesn't appeal to me, that whole thing of more, more, more, more, more for the sake of it . . . Growth isn't necessarily getting bigger, and bigger, and bigger. No, it's doing things in the right way, and creating the right environment. Fine-tuning."

In 2008, it was time to move; her rent at the 1,800-square-foot Mott Street space had quadrupled in ten years, and her shop-atelier had blossomed into a full-fledged company. While driving down leafy, cobblestoned Bleecker

Street, she spotted a For Rent sign on a turn-of-the-century building with windows on three sides. Perfect for her new studio, she thought. She took 1,500 square feet on the ground floor for her shop, and 6,000 square feet more, divided among the basement (for storage), the first floor (for wholesale, retail, and communications offices), and the second floor (for design, production, logistics, and finance).

On the hot summer day I dropped in, the second-floor workroom was slathered in sunlight and cluttered with metal rolling racks of swatches, paper patterns, muslin samples, and finished clothes. The production office was staffed with headphone-donning under-thirties, busy-busy on desktop computers. Across the hall, Mr. Huang and his design team were conducting fittings for the spring-summer collection. When we walked in, he was reviewing a pretty berry-red cotton velvet dress with an asymmetrical neckline.

Cornejo has almost all her ready-to-wear produced within New York city limits, making her one of about fifty brands to manufacture at least three-quarters of their output there. (She relies on leather artisans in Italy to make her shoes, sources some knitwear in China and Peru, and contracts specialty knitters in Bolivia to handcraft sweaters.) Most of her suppliers are in the Garment District—"Thirty-Sixth Street, Thirty-Eighth Street," she said. "When there is any drama at the factory, my people jump on the Six and go right up and check on it."

Since 2009, Cornejo has produced four collections a year—two for the runway during New York Fashion Week, two for showroom appointments only. She has two stores (New York and L.A.), plus online and wholesale, twenty-eight employees, and about $10 million a year in sales—the size Chanin is shooting for. It's a smidgen of the $5 to $10 *billion* that the mega-brands such as Dior, Gucci, Chanel, and Louis Vuitton pull in annually, but respectable for a one-woman, privately held company.

"There is a reason why Midtown Manhattan is New York's fashion hub," she said. "It's wonderful to be able to walk out of your office and go see your client."

THAT MIGHT NOT LAST FOREVER—not if Andrew Rosen, a third-generation garmento and, until recently, the CEO for Theory, has his way. For the last couple years, Rosen has been spearheading an initiative to relocate the entire Garment District to Sunset Park in Brooklyn. The borough already has a few independent centers. In 2012, former Ralph Lauren design assistant Bob Bland founded Manufacture New York, a fashion incubator established in the former Storehouse No. 2 of the US Navy Fleet Supply Base. Pratt Institute has the Brooklyn Fashion + Design Accelerator. And there is the Greenpoint Manufacturing and Design Center, set in a cluster of restored textile factories. All three have been frequented by members of Brooklyn's maker community.

During his State of the City address at the Apollo Theater in February 2017, Mayor Bill de Blasio announced his commitment to build a Made in New York campus in Brooklyn's Bush Terminal—just like the borough's president had unveiled twenty years earlier. This time, however, the city pledged $136 million for the venture, which would also include film and television production facilities. It's expected to open in 2020. "You're going to see this whole area come alive," de Blasio proclaimed.

In 2017, New York City had 1,568 apparel manufacturers, about one-fourth based in or around the Garment District. And not everyone is keen to move across the East River. That spring, designer Yeohlee Teng voiced alarm at a symposium on the subject: "We look at the district as an incubator." Joe Ferrara, president of the New York Garment Center Supplier Association, scorned the plan, calling it "a deportation."

Even those outside the fashion industry fretted. Like costume designer Steven Epstein: "If Bette Midler rips the train of her dress because the understudy chorus boy steps on it, the wardrobe supervisor can head to the garment center, buy the fabric, get it back to the theater, get that train recut and stitched before the show even begins."

Rosen dissents.

"I've been a big proponent of trying to reimagine what the Garment Center should look like twenty years from now, not what it looked like twenty years ago," he told me in his all-white, west-facing corner office at Theory's headquarters on Gansevoort Street in the Meatpacking District in 2017.

A six-foot-tall, sturdy man in his early sixties, he was digging into a plate of sashimi on a tray at his desk as he spoke. He stabbed the air with his chopsticks to reinforce important points.

"A lot of people are afraid of [the Brooklyn idea] because they think no one"—manufacturers, workers, clients—"will come there."

*Chopsticks.*

"But—"

*Chopsticks.*

"If we create a new vibrant community with state-of-the-art space and state-of-the-art equipment, people will want to work there, and it'll be a thriving modern industry."

In the early days, "everything" for Theory was made in New York City. "There was no reason to go overseas, because I could do everything here," he said, still tucking into his Japanese repast. "But as business diversified globally, there were a lot of advantages to manufacturing overseas price-wise." He reminded me about Liz Claiborne's trailblazing. "They were the first," he said, and "they changed the game. They built a multimillion-dollar company because they had the advantage of understanding how to manufacture overseas."

Today, he said, about a quarter of Theory's runs—"jackets, pants, tailored things"—are New York made.

The rest?

"All over. China. Vietnam. Peru."

Samples for Theory and Helmut Lang, another brand within the Link Theory group that Rosen ran, are designed, made, fitted, and shrink-tested in the company's sleek Design Center, a minifactory a block away on Gansevoort Street. Opened in 2016, the Design Center is outfitted with the newest technology, such as bonding machines—a system that seals seams instead of sewing them—and laser-guided pattern cutting, which allows for sharper precision and less waste. When a look is greenlighted, it's sent to a manufacturer in Midtown or overseas to be commercially produced. While Theory's hybrid model is not wholly rightshored, it's more so than competitors who make *everything*—samples included—offshore.

Would Rosen ever bring that offshored work back to New York?

"If it gets relocated and centralized in one place, it's possible," he said.

For small US-based businesses—those that do $10 million or less in turnover annually—producing domestically is a no-brainer. "It's quicker, you can keep your eyes on it," he said. "There are great advantages to manufacturing here, and I encourage young start-ups to do that, and do at least 75 percent of their manufacturing in New York. It's the best way to start a business. And it's more possible now than even five years ago."

He's certain that moving the Garment District to Sunset Park would make the whole endeavor that much more enticing.

"One of the problems in New York manufacturing now is that the pattern making is here, the sample making is there, the cutting is *there*." The chopsticks pointed east, then west, then south. "And the sewing is someplace else."

"I think everything can happen out there," he said, as he waved his left hand in the direction of Sunset Park—"because the younger side of this"—meaning the fashion business—"they all live in Brooklyn and can't afford to be in the city anyway. It could be amazing. You could have design, and manufacturing, and showrooms."

"Like a fashion city," I said.

"Yeah," he responded.

He pointed the chopsticks at me.

"Or a fashion ecosystem."

DIRECTLY BEHIND MARIA Cornejo's headquarters, on Bond Street, is a shop selling a lower-priced, West Coast version of her hyperlocal, ecologically minded fashion. Called Reformation, the brand was founded in 2009 by a Beverly Hills–raised former model named Yael Aflalo. Reformation is green. It is transparent. Rightshored. But Aflalo's ambition is much, *much* bigger than Cornejo's: she says she wants to be "a sustainable fast fashion brand."

She does not see this as an oxymoron.

Making clothes "quickly has no bearing on environmental impact," she told me.

Aflalo's recipe is simple: produce decent-quality fun fashion, such as capri pants, A-line mini dresses, and crop tops, in clean, aboveboard factories, primarily in greater Los Angeles, and sell them for a reasonable price: roughly $40 to $450. Fans include Taylor Swift, Rihanna, model Karlie Kloss (who became an investor), and Meghan Markle, the Duchess of Sussex, who donned Reformation's breezy $218 pale-gray-and-white-striped "Pineapple Dress" during the South Pacific royal tour in 2018.

"I design for this waitress girl," Aflalo explained to *Allure* in 2017. "She made $200 in tips the night before and she is obsessed with the dress. We're kind of trading the dress for those tips."

Aflalo borrowed the model from American Apparel's founder, Dov Charney. Before he was expelled by his board in 2014, following charges of alleged sexual harassment and mismanagement, Charney had proved that it was absolutely possible to produce low-priced clothes in the United States while compensating workers above minimum wage. As early as 2004, he was paying between $13 and $18 an hour and offering benefits such as healthcare and free English classes, and he still made a hefty profit.

I asked California Fashion Association head Ilse Metchek how Charney pulled it off.

"Economy of scale," she explained. "He did it on the premises. He didn't have to buy his fabric; he made it. He was vertical, totally vertical. And had very limited SKUs"—stock keeping units, or items for sale. "He had T-shirts, a hoodie. He didn't have a line of a hundred different items. He didn't have contractors; he made it himself. The machines were there. He didn't own property; he rented. People have a very different view of success. They want to own the building they're in. Then you say to yourself: Are you in real estate or are you in the business of apparel?"

Aflalo gets this. Her main factory, which she rents, is in the gritty, industrial suburb of Vernon—cheap. Much of the fabric she sources is deadstock from other companies—cheap. She has limited SKUs for her core—basic tops, jeans. But she dresses up her selection with frequent drops of short runs—yes, just like fast fashion. She believes if you cleave to the Zara business model, and analyze point-of-sale data immediately, reorder what is selling, and stop what is not—if you are agile—you'll sell more at full price, and you won't have leftovers. (Though fast-fashion brands do routinely have *tons* of leftovers.) And if you make a good-quality garment that costs more—and lasts longer—than traditional fast fashion, customers will think twice before tossing it. "Cheap is disposable," she said. "We try to make clothes that are not disposable. Our customers resell their clothes."

A glossy upspeak brunette in her early forties, with cheekbones you could hang the laundry on, Aflalo learned the trade as a youth: her parents had a fashion boutique in downtown L.A. In 1999, at twenty-one, she decided to give it a whirl herself and launched her own label, Ya-Ya—a play on her first name, Yael. The influential West Hollywood retailer Fred Segal ordered, as did a few other on-point specialty shops. By 2005, Ya-Ya was doing $20 million in sales. "I bought a big house, I had more than one car," she later said. "I had parties, and I would go on long design vacations." But Ya-Ya flopped during the 2008 recession. To pay her debts, Aflalo spent a year designing fast fashion for Urban Outfitters.

When she was financially stable again, she regrouped and started refashioning vintage dresses, which she sold under the name Reformation for a hefty markup at a Lower East Side boutique. In no time, business was booming, with stores in New York and Los Angeles. "I didn't work that much," she said. "I could afford the lifestyle I wanted to have."

She took a trip to China, was horrified by the waste and pollution she saw, and had an epiphany: from then on, Reformation would be green and socially aware—an approach she called "eco-chic." "I want altruism and narcissism to be combined," she proclaimed. She took over an old bakery in Boyle Heights—a low-income downtown neighborhood that has famously protested gentrification—and turned it into what she boldly claimed was "the first sustainable sewing factory in the US!" She even had a kitchen garden in the backyard, where she encouraged employees to grow their own vegetables. She bought a Tesla.

Aflalo has employed good practices, such as using sustainably sourced materials, posting the carbon and water footprints of the brand's products on its website, and attaching a "RefRecycling" label to items that allows clients to mail back clothes for repurposing. (Few have.) The company's slogan says it all: "We make killer clothes that don't kill the environment."

But while that was all well and good, her driving business principle, she said often and loudly, was something she learned at Urban Outfitters: "speed to market."

So devoted to this business doctrine is she that she hired Zara's trend director, Manuel Ruyman Santos Fdez, as her design director. Board member Ken Fox, founder of Stripes Group, one of Reformation's investors, told *Forbes*: "Yael has created that opportunity to be a next-generation Zara."

She's fine with that. In fact, she's gunning for that.

"There are things at Zara that are nice quality," she told me.

In her mind—and her business plan—there is no cap on growth. No such thing as too many clothes.

"The prevailing sustainable platform—'Buy less, use less,'—isn't a scalable strategy," she has publicly declared.

"I don't think it's a viable business option," she told me. "I think you'd have a hard time telling consumers to buy less and use less. There is a small percentage that might adopt that, but no, I don't think it's a realistic approach to climate change."

For her, sustainability means running a tight ship: using renewable or clean energy, like solar power—"We have fifty percent less carbon emissions than the industry average," she brags—and watching water consumption. Then "offsetting" the damage that is inflicted by improving the situation elsewhere. Like with water: "We work with nonprofits that clean waterways," she explained. "We are cleaning I don't know how many gallons of water across the United States." Such practices, she said, make her company carbon neutral, water neutral, and waste neutral.

In 2016, Aflalo moved her company from the Boyle Heights bakery—"We were exploding out of that place"—to the single-story 120,000-square-foot former headquarters of True Religion in Vernon after the jeans brand moved offshore. East of the 101 freeway, Vernon is a flat grid of warehouses, auto body shops, and clothing factories—legit and non—ringed by razor-topped chain-link fences.

Reformation's factory has no name on the façade. Like its neighbors, it is surrounded by a fence and parking lot—no more kitchen garden. About one-third of the company's output is cut and sewed there. T-shirts are produced in contracted factories in the area. Sweaters are made in China. The creative office, where Aflalo works, is about a twenty-to-forty-minute drive (depending on traffic) from the factory, in Platform, a posh new luxury retail and business center down the street from Sony Pictures Studios in Culver City. She has a shuttle bus that runs between her two outposts. She goes to Vernon herself about twice a month.

Like Charney, Aflalo adopted such admirable rightshoring practices as paying her plant workers above minimum wage and offering healthcare benefits, free massages on-site, and classes in career counseling, English language, and a path to citizenship. But don't get her wrong—she's not doing this because she's feeling patriotic. "I am not a diehard Made in the US person," she told me. "Our production is in the US rather than overseas because we are able to make it much faster here and sell it faster. We are selling more product more often." It's just smart business.

She also knows it's smart to show off her more humane business practices. On April 22, 2017—Earth Day—she flung open the Vernon factory's doors and invited the public to come see Reformation in action. The company announced it would continue the practice each first Friday of the month. I went on one of those tours in October 2017.

The reception space at Reformation's Vernon plant is bright and midcentury minimalist, with wiry Bertoia chairs and white Saarinen tulip end tables. The employees who strolled in were young and mod, and many had pooches—a bulldog, a rangy mutt, a white poodle named Elodie—on leashes.

"Is it Bring Your Dog to Work Day?" asked one of the thirty-five of us waiting for the visit.

"No," the receptionist responded. "You can bring your dog every day."

Our guide was Kathleen Talbot, vice president of operations and sustainability. In her early thirties, petite, and peaches-and-cream pretty, she was outfitted in skinny jeans, a year-old midnight-blue Reformation silk blouse, and beige flats with black tips—knockoffs of the classic Chanel ballerina shoe. Her stick-straight mahogany hair was pulled into a messy bun.

She told us that Aflalo's "vision" is "really to do something different." As Talbot spoke, we walked past conference rooms labeled "Chat Rooms" and the coffee nook, called "L'il Kitchen." The spacious "Chill Room" was outfitted with a long caramel leather club sofa, two big Pilates balls, a beanbag, and a stereo with an expansive collection of LPs. It may have been standard West Coast start-up culture, but it was a *long* way from Dhaka.

At the time, Reformation had eight stores—three in L.A., three in New

York, one in San Francisco, and a new one in Dallas. The clothes were designed by four creatives—two designers (one being Aflalo) and two assistants. A force of four hundred worked for the company in total; three hundred of those people were in this factory. We were "welcome to take photos, videos," Talbot said. "There is nothing off-limits."

Our first stop was production development—an enormous, double-height, all-white room with bolts of fabric, and cones of thread, and women sewing on industrial machines.

"We are actually 'fast fashion,' which I know sounds like a dirty word," Talbot said. "What we mean by that is that we make clothes really quickly, and we're very responsive to the trends and the demands of right now. We actually release twenty to fifty styles a week. We work on a weekly basis, not traditional quarters or seasons. We can go from a design concept to shipping to customers in as little as four weeks. Our average is forty-two days. And if we make something, and it does very well, we can make more of it in as little as two weeks. What that lets us do is always be on the pulse. It also lets us make *less*. [We] make really small runs and sell it all, instead of having extra inventory. The discounting that you see in traditional fashion retailers is mistakes. They bought too much, they made too much, and they're trying to just offload. We really are a little more nimble."

We saw the Fabric Room, with more sewers and more bolts of cloth. And the Vintage Room, where used clothes were retailored, silk screened, washed, and dyed to sell at Reformation's new "Melrose Vintage" shop. And the Denim Lab—denim was a new line for Reformation. Aflalo believed sales would top $140 million in 2018, 80 percent of which would be e-commerce, and 20 percent in stores. When we spoke that September, a year after my tour, she had just opened her thirteenth boutique, in the Georgetown neighborhood of Washington, DC. She had also stepped into wholesale, at Nordstrom. Plans were afoot to roll out children's clothing, menswear, handbags and shoes, and "one hundred cool stores," she said.

Eventually, Aflalo told me, she wanted to take Reformation public. In May 2019, however, Goldman Sachs was reportedly brought in to sell the company. Unlike when Imogene + Willie moved to L.A. and tried to scale up, Aflalo has cleverly enlisted a slew of business-savvy backers and advisers to help her achieve all those goals.

"Yael has lofty aspirations," Andrew Rosen, who is one of her investors, told me. "But if someone can make it happen, she can."

# SIX

# My Blue

# Heaven

MUDDY BOOTS were lined up outside the front door of Stony Creek Colors's headquarters in a one-story strip building in Goodlettsville, a country town about fifteen miles north of Nashville. "We were out in the fields this morning," farmer and entrepreneur Sarah Bellos explained, as she padded over to her desk—a barn door on trestles—in stocking feet. Thirty-three, fresh-faced and light-eyed, with her straight chestnut hair in a casual topknot, she was dressed in a gray T-shirt and jeans, by San Francisco indie brand Gustin, made of denim dyed with her indigo. Its cornflower blue was as vivid and luminous as a sapphire. Her fingernails were black from indigo dye.

"The fields" were thirty-five acres that she rented in nearby Greenbrier, where she contracted ten local farmers to plant and nurture the indigo; she helps plant, regularly checks on progress, is on hand for the harvest, and converts the leaves into dye, which she sells to manufacturers. Indigo has not been farmed commercially in the United States for more than a century—not since synthetic wiped out the natural indigo trade. (It is still farmed a bit in India and Japan.) Bellos wants to change that.

Almost all the denim we wear—99.99 percent—is dyed with synthetic indigo. What is not publicized—what the apparel industry discreetly sidesteps—is that synthetic indigo is "made of ten chemicals—including petroleum, benzene, cyanide, formaldehyde—that are toxic or harmful to humans," Bellos said.

The denim industry continues to rely on synthetic indigo for the same reason fashion companies have clothes made in sweatshops.

"The economics aren't there for people to care," Bellos told me. "You can buy benzene for cheaper than paying a farmer to grow crops. Pollution is the cheapest way to do business."

With Stony Creek Colors, Bellos wants "to prove a more sustainable model could exist and be profitable."

For someone set upon disrupting the denim industry and changing the makeup of what we all wear, Bellos's background is decidedly un-fashion. She grew up in Mt. Sinai, Long Island, "in a house from the 1700s in the woods," she told me. "I played in the wetlands or in the woods. So I've always been captivated by the environment." Her mother was an artist. Her father studied forestry and worked as a carpenter.

She attended Cornell University, where she studied natural resource management. She interned at North Carolina State University—Natalie Chanin's alma mater. Her first job was with an investment research group in Washington, DC, that specialized in corporate social responsibility. But, she said, "I was not meant to be a research analyst. I wanted something more tangible and hands-on. I believe it takes smaller companies to prove it's possible before bigger companies will take the leap."

She quit and moved to Nashville, where her older sister, Alesandra, lived and worked as an artist. They decided to launch their own sustainable small business: Artisan Natural Dyeworks, a dye house that collaborated with independent designers, including Chanin. They learned the dye process "through books and a lot of trial and error," Bellos said. As pleasing and pro-green as it was, "hand-dyeing wasn't going to move the needle on what I saw as the sustainability crisis in the industry," she said. So, in 2012, she founded Stony Creek Colors, named for her four-and-a-half-acre farm in Whites Creek, on the edge of Nashville. With it, she wanted to "take the best com-ponents" of custom dye work, like "artisanship and transparency," she said, "and show it can be done on a larger scale."

When I went to see Bellos, in the summer of 2016, only a handful of arti-sanal indigo dye companies existed in the United States, and they were small and "more about maintaining the tradition," she said. She was the only one growing indigo on an industrial scale and supplying a major denim mill: Cone Mills's White Oak plant in Greensboro, North Carolina.

She landed that contract through luck and determination. In 2013, she met Neil Bell, then Levi's chief innovation officer, while on a sustainable cotton tour in California. When she told him about her work, he was astonished. How could natural indigo be grown and produced commercially in the United States after so many years of total reliance on synthetic? She assured him it could be—that she was doing it. He took note.

The next year, Bell introduced her to longtime Levi's partner Cone. At the time, Cone was one of America's four remaining industrial denim mills. "Crazy," she remarked, given that "denim was invented here."

She told the Cone executives that she wanted to bring alternative crops to farmers in the southeastern United States and to reshore textile manufactur-ing. "Both parts of that fit with Cone's vision of authenticity," she said. Her first batch for them, in 2015—for up to 50,000 pairs of jeans—went so well, the mill precommitted for 2016 to enough indigo to dye up to 150,000 pairs. With that contract, Bellos was able to raise financing to open a proper fac-tory, as well as contract an additional fifty acres for farming. She is the ma-jority shareholder of Stony Creek Colors, but she also has individual and institutional investors.

She walked me back into the lab behind her office. With its collection of stainless-steel tanks, it looked like a microbrewery. She and her team stew the indigo in the vats, and when they are happy with it, they reduce it to a paste or powder, which can be stored and sold to manufacturers. Her two chemists test batches of indigo to see "how many pounds of fabric—and thus pairs of jeans—you can dye with a powder dose." Or what sort of shade they'd get if they adjusted the pH or other dye conditions—if it will be "more purply blue" or "more blue-blue," she said.

She also works with black walnut, which generates deep to muted browns; osage, a hardwood from the South, which gives her a palette of oranges, olive greens, and vibrant yellows; and madder, an herbaceous perennial that creates luscious shades of cranberry red. But indigo is her main gig: it accounts for 95 percent of her production and most of her $1 million annual revenue.

She concedes that, at first glance, natural dyes appear to be more expensive than those used for petroleum-derived synthetics, simply because of the higher cost to farm raw materials. But there are "externalities" not included in synthetics' calculation—environmentally damaging ones "like fracking and oil spills," she said. The false economy that so disturbed Cornejo when she worked for a fashion multinational has a parallel in the dye industry.

There are other concerns with synthetic dye. As Bellos explained it, a handful of Chinese manufacturers produce most synthetic indigo today, and for it they use a chemical called aniline. The EPA classifies aniline as a Group B2, probable human carcinogen, meaning it may cause cancer. And the US Centers for Disease Control and Prevention has declared it "very toxic" to aquatic life. Recent reports indicate that two-thirds of aniline residue winds up in wastewater, lakes, rivers, and other waterways; on workers; and in the air that workers breathe. The remaining third is embedded in the denim jeans, jackets, and skirts sold in stores.

She suggested we go visit the fields in Greenbrier, a fifteen-mile drive from her office. They were indeed muddy. Eleven of the thirty-five acres were planted. The soil was clay-like and somewhat sandy—a mixed loam. It wasn't organic; since she doesn't own the fields, she can't control what's done when she's not farming them. But her operation does not use pesticides.

Bellos's farmers grow three varieties of indigo: one from Japan (*Persicaria tinctoria*) and two tropical (*Indigofera suffruticosa* and *Indigofera tinctoria*). The Japanese plants were hip high, dark green, and bushy, like giant basil. The air smelled pungent, sweet, and acrid. The tint, she explained, is in the leaves. She plucked a couple and handed them to me. I stuck them in my notebook; as they dried, they turned the same blue-black as her fingernails. Some of the Japanese plants were flowering, the blooms a piquant magenta. Butterflies, ladybugs, and red wasps, which eat other bugs, swarmed. "Honeybees love it, too," she said. Generally, pests find indigo bitter and leave it alone. "Insects don't come back if it doesn't taste good!"

On the next field were the two tropical varieties. One was taller—up to my belly button—and gangly. It's a perennial, but the stem can turn woody, so she grows it as an annual. The second was shorter and small leafed. Its weight yield is far less than the other two, but its potency is fierce. "Like concentrate," she said. Some of the plants she lets go to seed, for the following season.

Tennessee has long been a tobacco state. But with cigarette smoking on the wane in the US, tobacco farming has declined. Thankfully so, from the farmer's point of view. Tobacco is finicky and labor-intensive—planting and harvesting are still done by hand. "It's a nasty job. Every bit of it is hot and nasty," Robertson County farmer Larry Williams said. "None of the young people want to work in tobacco . . . Can't blame 'em."

Susceptible to disease, tobacco requires copious amounts of costly herbicides and fungicides. And it hungrily leaches nutrients from the soil and requires up to "three hundred pounds of nitrogen fertilizer per acre," Bellos explained. Indigo—a legume, like soy—needs far less feeding: while one variety that Bellos grows (*Persicaria tinctoria*) calls for 70 pounds of supplemental nitrogen per acre, the two tropical varieties require zero; they actually infuse the earth with nitrogen.

Bellos had no trouble getting farmers to participate and sign on. She pays them 20 to 45 percent what they earned growing tobacco—$1,000 to $1,800, rather than $4,000 to $5,000. But their costs are so much lower, they pocket roughly $550—instead of $87—per acre. To make the work more efficient, she has developed a mechanized indigo harvester, which members of her staff run. Indigo, Bellos said, gives farmers "a viable economic alternative" to traditional crops. It's "revitalizing rural economies," she said, as well as "nourishing the soil."

We drove about fifteen minutes to Springfield, the Robertson County seat and home to her "factory"—an eighty-thousand-square-foot redbrick warehouse, built in the 1950s. Upon entering it, I was knocked sideways by a sharp scent that burned my nostrils. For more than half a century, the Conwood warehouse processed dark-fired tobacco for chewing and snuff, and the porous bricks absorbed much nicotine. As a nonsmoker, inhaling it gave me a slight buzz.

When American smoking rates dropped and the tobacco business moved offshore—yes, tobacco, too—the factory's last owner, the American Snuff Company, a Reynolds American brand, shuttered it. In 2015, Reynolds donated the factory to the county. The closing "was an economic disaster for the area," Bellos said.

She decided to repurpose the building—to rightshore, in a sense. It sits directly across the road from the local water treatment plant, so Bellos can easily dispose of her processing wastewater. And the county rents to her for an attractive rate. "We have room to grow here!" she said with a laugh as we took in its dark, empty immensity.

We walked downstairs to the parking lot out back. There were bales of indigo piled up and an enormous Dumpster-like container—called an extractor—filled with water, some of it from her cistern. Bellos can collect up to one and a half million gallons of rainwater a year, captured off the roof. The indigo leaves are loaded into the extractor and soak for a few hours.

The inky liquid is pumped into a six-thousand-gallon stainless-steel vat—the industrial version of the microbrewery—where it is mixed with oxygen and transformed into dye. The soup is then transferred into settling tanks, where it separates; as indigo is an insoluble pigment, it sinks to the bottom. The top layer of water is filtered and reused or sent to the waste treatment plant. Bellos showed me a plastic tub filled with processed indigo juice. It smelled like cat pee. When I said this, she laughed. "In medieval Europe, they used urine to reduce the indigo dye and make it soluble." The pigment is transformed to paste or powder. Cone preferred paste, which Bellos had trucked to the White Oak plant.

Bellos had big plans for growth. For 2017, she aimed to plant 165 acres— a threefold increase—and by 2021, she wanted to be up to 17,000 acres. "We are producing nowhere near what the market wants," she said. "It's going to evolve, especially with customer awareness of such things like the use of cyanide in your jeans."

Bellos told me in 2016 that if she met her goals, she hoped to command 1 percent of the entire indigo market by 2021. Two years later, that target figure had grown to 2.8 percent by 2024.

But she wants to do more than that.

If she has her way, "there will be more companies like us in ten years," she told me. "We'll see."

SLOW FASHION has infiltrated jeans production, too.

A bevy of boutique denim brands, such as Blue Delta Jeans Co. in Oxford, Mississippi, and Hiut Denim Company in Wales, have cropped up, crafting small runs of artisanal jeans. "Our job is to make the best jeans we can, not the most jeans that we can," Hiut cofounder David Hieatt explained. He got the top style endorsement of the moment when, in early 2018, royal fiancée

Meghan Markle wore a pair of Hiut jeans to an official visit in Cardiff; overnight, the company was overrun with orders, creating a months-long waitlist.

But in terms of both craft and mystique, Japanese denim makers are beyond them—*way* beyond.

For denimheads—a tribe so dedicated to the blue cloth their obsession borders on fanaticism—selvedge jeans from Japan are the *ne plus ultra*.

Best of all are jeans made in Kojima, a small town four hours southwest of Tokyo by train, in Okayama Prefecture. So respected is Kojima's denim trade that every time I have mentioned it to anybody in the denim business, everyone—and I mean everyone—has flat-out swooned and professed that it was their *dream* to go see it in person.

Kojima was poised for textile success: the town was built on land reclaimed from the sea and therefore had poor soil. All they could grow abundantly was cotton, and like in the American South, mills sprang up to spin and weave it. Kojima evolved into a manufacturing center in the mid-twentieth century, specializing in school uniforms made of cotton cloth. By the 1960s, the town produced 85 percent of the country's student blouses, blazers, pants, and pleated skirts.

In 1965, Kotaro Ozaki, of Maruo Clothing, decided to steer his company into a new domain for Japan: blue jeans. But not just any sort of blue jeans. These would be knockoffs of American jeans, made of denim from Georgia's renowned Canton Textile Mills (which closed in 1981). Japanese passion for American blue jeans dated to the 1940s, when American GIs stationed in Asia wore them on leave in Tokyo. The Brando-Dean-Elvis look of the 1950s reinforced jeans' cool in Japan. With a cultural obsession for quality, the Japanese considered American-woven denim to be the gold standard.

Two years after Ozaki's first Canton-sourced jeans hit the local market, he introduced a label called Big John. For it, he bought the best denim of all: Cone White Oak. In 1970 came a women's line, Betty Smith—a name Ozaki also dreamed up to sound as American as possible. Maruo jeans looked

American, and they were made of American denim, but they were strictly for the Japanese.

In 1972, Kurabo, a mill in nearby Kurashiki, began weaving denim—another first for Japan—which kicked off a veritable jeans mania. More brands cropped up, all with American-ish names, like Bison and Big Stone. Sales of Japanese-made jeans jumped from seven million pairs in 1969 to forty-five million in 1973. Folk singer John Denver starred in—and wrote the jingle for—a Big John commercial. Soon, 70 percent of all Japanese jeans came from Kojima—the "Holy Land of Jeans."

As much as the Japanese loved their American knockoffs, they also craved the originals. They believed US denim was the best quality—the top being Cone's White Oak selvedge—and they had a reverence for what the garments had been through: the life, the experience, the breaking in that had been so hard-earned.

In the 1980s, Japanese "pickers" trawled flea markets and secondhand shops in the United States, hunting for old Levi's, Lees, and Wranglers, usually tagged for less than $20 apiece. The jeans would be resold at an eye-popping price in Japan—sometimes as much as fifty times what the picker had shelled out stateside.

The vintage jeans weren't just for wearing; they inspired new Japanese versions. In Osaka, five boutique manufacturers began putting out top-of-the-line interpretations of the American classics, made of hand-dyed selvedge denim woven on twenty-seven-inch looms and finished with studiously sought details, like cinch-back buckles from France. One company went so far as to claim it used antique looms purchased from White Oak, though experts have since dismissed that as "pure storytelling." The Osaka Five reproduced the seasoned originals—particularly 501s—so precisely it was difficult to tell the old from the new. Osaka-made "new vintage" jeans had a serious cult following.

Until Kojima stepped up.

Today, a bus with denim seats ferries tourists from the train station to the center of town, where you can visit a jeans museum and stroll down "Jeans Street": a pedestrian walk lined with Americana-decor boutiques offering local denim wear and snack bars hawking "blue jean" ice cream—a cyan concoction that tasted like Pixy Stix to me.

Kojima's star brand is Momotaro Jeans, an artisanal line of selvedge jeans introduced in 2006 by the local denim mill Collect Co. and named for the Japanese fairy tale about a boy born from a peach and raised by a childless couple. The brand's logo is a charming illustration of a pudgy lad with a black bowl-cut bursting out of a fleshy drupe. Momotaro only produces forty-five thousand pairs of its machine-woven, natural-indigo-dyed selvedge jeans a year. They sell for around $300 a pair, in selected international shops and online, and are the sort that make denimheads gaga.

On a misty spring morning in 2018, Momotaro's general manager, Tatsushi Tabuchi, an affable thirty-three-year-old sporting the denimhead uniform—plain white T-shirt, selvedge jeans (in his case, Momotaros) cuffed to reveal the seam, and black work boots—took me to see the company's factory, on the edge of Kojima. Momotaro's denim is woven on nine forty-year-old Toyoda shuttle looms—old workhorses developed by the Japanese textile industrialist Sakichi Toyoda, father of the founder of Toyota Motor Corporation. The looms sit on wood-plank floors, to absorb vibration and allow for movement, and the roar is deafening. "The production of these machines has been discontinued," Tabuchi shouted at me. "We keep old ones in the back for parts."

The looms each put out roughly fifty meters of selvedge a day. The cotton comes from Zimbabwe. Most of the indigo is farmed and fermented in Okinawa or Tokushima, and is "the highest quality," Tabuchi said. I massaged the cloth between my fingertips—as Mary Katrantzou's fabric developer, Raffaella Mandriota, had taught me to do at Première Vision. The surface was coarse, and it had a few flaws.

"With these vintage looms, we can weave the denim more unevenly, which gives it an interesting face when worn over time," Tabuchi explained as we walked out of the weaving room. "The craftsmen advise us on which machine to use to get the right effect, because each has its own personality. The looms are old, and fussy, but we can change the thickness of the fabric, the number of wefts, and the tension of the warp—delicate adjustments that give a subtle change to the cloth. That is why Japanese denim is so prized."

In the sewing room, I caught sight of a couple of seventy-year-old black Union Special sewing machines for hemming—scooped up from Levi's US factories closed down by Marineau. There was also a more recent beige Union Special—1970s or so—to sew hip seams and crotches with the chain stitch. "When the jeans are washed, the chain stitch shrinks, like the denim— that's the advantage," Tabuchi said. He added that he had "good mechanics" to repair all the Union Specials when needed. These jeans, you could say in fashion-speak, were Momotaro's "ready-to-wear" offerings.

In 2006, Momotaro introduced its version of couture: jeans made to order by hand with denim woven on an antique manual kimono loom from Kyoto.

Tabuchi told me it was Momotaro's response to the apparel industry's blatant overproduction of jeans. Of the six-billion annual output, exactly twenty are Momotaro's kimono-loom-woven, handmade gems. "We wanted to create the ultimate jeans," Tabuchi explained. "Japan has a tradition of handweaving," and working on the kimono loom "preserves that technique."

We headed to the Momotaro shop on Jeans Street, where the loom is located, to watch a demonstration. Kazuki Ikeda, a handsome twenty-six-year-old man in a tailored denim suit (neatly folded white handkerchief in the jacket breast pocket), blue shirt with French cuffs, blue necktie, and polished shoes, stood adroitly at the loom, pushed the beater forward, slid the shuttle through, then pulled the beater back toward him: *thunk, thunk*. He pushed it forward again, zipped through the shuttle, pulled it back: *thunk, thunk*. He had trained for five years with a master to learn the technique. To get the weave right, and maintain consistent strength and tension, he must apply the same pressure with each movement. During an eight-hour workday, he can weave seventy centimeters of denim. A bolt is fifty meters. "The selvedge looms are slow," Tabuchi whispered to me as we watched. "But the kimono loom is even slower."

The superior quality is obvious to the eye and to the hand. The surface of the thread is round—mass-production machines crush that roundness—and there is a fluffy feel to it. It takes three months to weave one bolt of cloth—thus the twenty-pairs-of-jeans-a-year run. Despite their 200,000 yen, or roughly $2,000, price tag, there is a wait list. And in mid-2016, Momotaro stopped taking orders, because, Tabuchi said, "We couldn't keep up."

THE JAPANESE PROVED that high-quality, profitable, and desirable jeans can be made following the slow-fashion model.

But demand for high-end selvedge denim is relatively small. Distressed, whiskered jeans continue to drive the global market and have been the cause of ecological and health calamity. What could be done about finishing? Surely, in our world of rapid technological advancement, there must be a way to banish the horrors I saw in the Ho Chi Minh sweatshop?

Denim industry consultants José Vidal and his nephew Enrique Silla, based in Valencia, Spain, posed themselves the same question more than twenty years ago and set to developing a cleaner, safer process.

Called Jeanologia, it is a three-step system: lasers, which replace sandblasting, hand-sanding, and the bleaching chemical potassium permanganate (PP); ozone, which fades fabrics without chemicals; and e-Flow, a washing system that uses microscopic "nanobubbles" and cuts water usage by 90 percent. Some clients subsume one Jeanologia step into their jeans finishing structure; some two; some all three. Every step makes a difference for the better.

Traditionally, finishing a pair of jeans requires an average of 70 liters (or 18 gallons) of water, 1.5 kilowatts of energy, and 150 grams (or 5 ounces) of chemicals. In total, that equals an astonishing 350 million cubic liters (92 million gallons) of water, 7.5 billion kilowatts of energy (enough to power the city of Munich for a year), and 750,000 tons of chemicals *each year*, ad nauseum.

The Jeanologia system can decrease energy consumption by 33 percent, chemicals by 67 percent, and if implemented most efficiently, water usage by 71 percent—or, as the company proudly boasts, to one glass of water per pair of jeans.

"Jeanologia was born to completely transform the way we produce textiles," Enrique Silla, the company's chief executive, told me in his plain office at the company's headquarters, a 1990s corporate building outside the city center. "Our mission from Day One was to eliminate contamination, to take care of people, and to take care of ecology."

An elegant man in his fifties, with pewter hair combed back from the temples, Silla was dressed neat-casual in a well-pressed chambray shirt and, of course, jeans. He runs the company with his sister, Carmen Silla, who serves as brand and marketing director. Along with the Valencia campus, there is a laser production factory in Barcelona and a service and development center in Izmir. (Turkey is another major denim production locus in the fractured apparel supply chain.) Of the six billion pairs of jeans manufactured and finished in 2018, about 30 percent went through at least some part of the Jeanologia system.

Silla led me to the lab to see the system in action. In the laser room, a young man pulled virgin jeans onto the legs of a half-mannequin in a clear-walled cabin. His partner stood nearby at a computer control panel and launched the process: the lasers descended the face of the jeans—I assumed that was what was happening, since the beams were invisible—and clouds of blue smoke puffed as distress patterns emerged. In ten or eleven seconds, it was all over: the jeans were as faded and destroyed as my old shrink-to-fit 501s after three years of hard living. "Masks and pollution are a thing of the past," Silla said. "Technology is the way forward. It's cleaner and healthier."

In the next room, he showed me a dryer-like tumbler called the G2 Cube that uses ozone to fade jeans. Stratospheric ozone, or "good ozone," is a natural gas found in our atmosphere (versus tropospheric ozone, or "bad ozone," which does not exist naturally but is born through human actions, like car emissions). Using good ozone in finishing is "like putting a garment in the sun for a month, except we can do it in twenty minutes," Silla explained, and with a fraction of the energy or water the old process required.

Finally, we visited the washroom, outfitted with e-Flow: a machine that washes jeans among bubbles so microscopic, they tot up to one million per cubic centimeter. With the traditional wash system, you saturate the entire garment. But the nanobubbles are like a foggy day in London—they only penetrate the surface. I stuck my hand in the machine. The sensation is a delicate damp—like a steam bath, but cold—with a bit of pressure and a little prickling. "Nanobubbles do the softening, tinting, and stonewash without the stones, all at once," Silla said. There is no water treatment afterward and the bit of water that is used can be recycled for thirty days. "We are not at zero water stage *yet*," he said. "But we are getting there."

While in Ho Chi Minh City, I toured a Jeanologia-equipped washhouse to see the process on a commercial scale. What a difference from the sweatshop I had visited earlier in the day. The modern redbrick facility sat in a walled compound replete with fountain and fishpond at the entrance. Inside, the spacious glass-partitioned rooms were bright, immaculate, quiet, and air-conditioned: the machines, as well as the workers, needed to be housed in a steady cool temperature to avoid overheating. Jeanologia, my host told me, had "totally transformed production."

I saw what he meant. In the e-Flow washroom, the floor wasn't flooded, workers didn't need Wellies, and their hands weren't blue. In the dry process room, whiskering was achieved by ozone, not chemicals. In the distressing rooms, lasers moved invisibly down the jeans, the wisp of fine blue dust sucked up by the vacuum system. The factory processed twenty-five thousand to thirty thousand pairs a day—about half of what the big Chinese factories churn out. There was no frenetic grabbing and flinging of garments, no screeching sanders, no sweltering heat, no stress.

I asked about job loss—an argument I hear all the time from opponents of manufacturing automation. "Eventually everything will be robotic," my guide conceded. But instead of waves of layoffs, this factory retrains workers to use "more sophisticated machinery, or to manage," he said. Again, like I'd heard at English Fine Cottons in Manchester, and about the rightshoring movement in the Carolinas, while there were fewer jobs, they were safer, more hygienic, less physically taxing, and better paying. These folks would not be hurriedly hand-sanding jeans and inhaling fibers and synthetic indigo dust as they labored. Back in Valencia, Jeanologia has a school to teach workers how to run its laser distressing machines. After the four-month course, newly minted "laser design experts" are dispatched throughout the world. "A cleaner industry," Silla told me proudly.

Even with all these pluses, Jeanologia has had a tough time breaking into the jeans finishing market. Initially, the chemical industry saw the process as a threat—because fewer chemicals means fewer chemicals. But Silla and his team proved that clean business is good business, and the large chemical producers came to appreciate and support Jeanologia's initiative.

Harder has been changing "the traditional guy who has been washing jeans the same way forever," Silla said. But they are making inroads "because of cost. Being efficient makes money." Jeanologia has a stand at Première Vision to help spread the word.

What would really change the game, however, would be to win over the majors: Gap, H&M, Zara, Uniqlo, PVH, VF Corp, and Levi's.

"If we transform the way these people produce," he said, "that would be *immense.*"

WITH LEVI'S, there was hope.

After decades of greedy, careless, and frankly irresponsible decisions that shattered small towns and nearly bankrupted the company, in 2011, Levi's made what may turn out to be its smartest executive hire in more than a century: Chip Bergh, a lanky, fifty-three-year-old former US Army captain who had spent twenty-eight years at Procter & Gamble, the last few as group president for global male grooming.

His mandate: set the company back on course.

"Keynesian economics talks about creative destruction, and it's a painful process," Paul Dillinger, head of global product innovation at Levi Strauss & Co., told me. But it also allows you to "come to the table with a fresh point of view, or optimism tempered by pragmatism."

And that's what Bergh brought to the table at Levi's.

A vegan who rises at five a.m. each workday, runs marathons, and competes in triathlons, Bergh promised he would "change the culture" at Levi's. The days of brand-tarnishing decisions taken for short-term gains were *finito*. He replaced ten of Levi's eleven top executives and two-thirds of the US-based managers and vice presidents. One of those recruits was James Curleigh, who served as brand president from 2012 to 2018.

"JC," as he likes to be known, is the most rock-n-roll executive I've ever met.

Six foot four and built like a lumberjack, with salt-and-pepper mad-scientist hair, a graying scruff, and a chin you could take a swing at, Curleigh is one of a set of identical twins born to a Canadian general and a mother whom he describes as a "free spirit." He made his bones as president and chief executive of Salomon Sports North America; when Levi's recruited him in 2012, he was president and CEO of KEEN, Inc., a footwear company in Portland, Oregon. His job was to make Levi's cool again.

Curleigh figured out how as his flight approached the San Francisco International airport. "[I could] see Levi's Plaza on the Embarcadero, and Silicon Valley," he recalled. "And I thought: 'What if we could take the iconic past of Levi's and . . . this modern-day entrepreneurial energy known as the start-up culture and create a positive collision?' . . . 'How do we become a one-hundred-fifty-year-old start-up?'"

Simple, he thought:

"We need our own garage."

Bergh had been thinking the same thing. And was already blueprinting one.

At the time, Levi's research and development center, run by denim expert Bart Sights, was located at the company's production center in Çorlu, Turkey, about seventy miles west of Istanbul—yes, Levi's had offshored innovation, too. It was a ridiculous setup that embodied the fashion's false economy: whenever the design team in San Francisco wanted to try something new, or the lab techs in Çorlu wanted to show managers what they'd come up with, someone would have to hop on a plane or FedEx a sample halfway around the world. "We probably spent enough on airfare to buy a 747," Bergh quipped.

In 2012, he decided it was time to bring innovation back home to San Francisco. With Sights and chief supply chain officer David Love, he scanned the city for potential locations and, after twenty or so, arrived on the old Eureka Grain Mill, a two-story, nineteenth-century redbrick building at the foot of Telegraph Hill that had most recently been the seat of a tech company. "As soon as we walked in, we knew this was it," Sights told me. "All we did was rip out the cubicles."

We were standing on the ground level of the lab, a wide-open space framed by a first-floor gallery. Small and fiftyish, with a shaved head, serious biceps, and translucent blue eyes behind NASA-scientist glasses, Sights was dressed in a white T-shirt, cuffed selvedge denim jeans, and black work boots. His fingernails were black, like Sarah Bellos's.

He was raised in Henderson, Kentucky, and began his career at the family firm, Sights Denim Systems, one of America's first large-scale jeans finishing plants; at its apex, its seven hundred employees washed and distressed jeans for such brands as OshKosh B'gosh, Lee, Wrangler, Gap, and Levi's. (His sister, Imogene + Willie cofounder Carrie Eddmenson, learned the trade there, too.) When it closed in 2008, a victim of the offshoring exodus, Sights headed to India—another major jeans manufacturing center—to work in design and development for Raymond UCO mill. Two years later, under John Anderson's leadership, Levi's hired Sights to oversee the Turkey R&D center. And three years after that, he created the Eureka Innovation Lab—the "garage" Curleigh had imagined.

Today at the lab, Sights and his team of thirty technicians "test jeans for strength, stretch and recovery, durability, water repellency, how the garment would react in a home laundry—all sorts of things," he explained to me. On the cement floor lay a dozen pairs of Levi's in various states of distress, scrutinized by several young, jeans-clad technicians. "We do the initial prototypes—we'll do a thousand pant finishes a season—and everything we create has to go through the filters of scalability," he said. "We need to be able to ensure that if we make one, we can make a million of them."

Across the room was a fully outfitted minifactory with pegboards holding big cone-shaped spools of thread in Levi's basic colors of navy, black, white, and goldenrod; bolts of denim leaning against a wall; and nine seamstresses at machines, running up jeans for Sights's team to test.

The lab didn't "operate like a traditional sewing room, with the cutting tables a mile long, spread with fabric, and you cut through, and send those bundles to operators, who join everything together, and do the same thing over and over," Sights said as we looked on. "Our sewing operator will do everything from start to finish, and do the fittings, and work with the pattern maker, and oversee the many iterations of a design."

When Levi's advertised to fill those jobs, a covey of former workers from the Valencia Street factory, which had closed in 2002 after a ninety-six-year-long run, applied in person.

"Those are our jobs," they said.

They now account for eight of the lab's nine sewing operators.

Upstairs, where the floor is sisaled, the colorful chairs are by Eames, and a rainbow flag hangs from the balcony railing, is the atelier for chief tailor Rachel Keene. On that day, she was dressed in a crisp white oxford-cloth shirt and high-waist selvedge jeans with suspenders, her kelly-green-streaked blond hair twisted up in a chignon held by a pencil. She was working on prototypes and special orders.

And across the mezzanine from her is Levi's Vintage Clothing (or LVC): a capsule collection rooted in the company's history rather than trends, designed by Paul O'Neill. A Dubliner, O'Neill joined the LVC team in 2009, when it was based in Amsterdam—more offshoring, which Bergh also called back home. For inspiration, he delves into the Vault and conjures cool Americana looks, like 501s in selvedge denim with a high waist and suspender buttons, or 1970s-style chambray shirts with pearly snap closures. O'Neill told me about one of his pet projects, a new line of jeans made of selvedge denim woven by Cone's White Oak plant and dyed with American-grown natural indigo.

Sarah Bellos's indigo.

Beyond sewing is finishing. The necessary tools are stationed throughout the Eureka Innovation Lab, including industrial washers and dryers and vats of natural indigo. In a small back room called "the physical testing lab," there is a rack from which strips of denim are weighted to observe the fabric's elasticity, and a microwave-like box in which swatches spin around on a mini merry-go-round, to measure ozone impact. Ozone, Sights explained, is the "strongest oxidizer on Earth."

In the back corner, I spied a Jeanologia machine. At the time, the lab was testing the system. Four months later, Levi's announced the launch of Project F.L.X., which stands for "future-led execution." It is an "operating model that ushers denim finishing into the digital era . . . by replacing manual techniques and automating the jeans finishing process," the company stated. (That was not the only news: in March 2019, Levi's launched an IPO.) Among the tech innovations Project F.L.X. was introducing broadly throughout the Levi's supply chain: Jeanologia's laser distressing systems. In March 2019, Silla told me that Uniqlo had also announced it had brought on Jeanologia's full process, PVH had adopted the zero-discharge technology, and VF Corporation would be soon shifting to laser distressing at its Mexican facilities.

"This," Bergh said, "is the future of jeans."

ONE WEEK AFTER Paul O'Neill excitedly showed me the Stony Creek–
dyed White Oak selvedge he was using for his next Levi's Vintage Collection, Cone abruptly announced that it was closing the mill on Dec. 31. After
more than one hundred years, it had ten weeks left.

White Oak had been wobbly for a while. In 2003, Cone Mills Corporation filed for Chapter 11 bankruptcy protection. The following year, the
company's assets were acquired by WL Ross & Co.—a private equity firm
owned by investment banker Wilbur Ross that specialized in distressed
assets—for $46 million. It was folded together with Burlington Industries to
create a new company called International Textile Group (ITG). During this
period, industry sources told me, Cone stopped doing business as it had since
day one. "They kept outsourcing and kept outsourcing," one southern mill
owner told me. "And they spun less and less and less. And bought other
yarn." And didn't update systems—they were stuck with twentieth-century
inefficiencies. Sales dropped, from $900 million in 2005 to $610 million in
2015. White Oak was the last major American producer of selvedge denim,
but the coveted cloth accounted for a smidgen of the mill's overall output;
less than 10 percent. Not enough to keep the place afloat.

In October 2016—weeks before Donald Trump was elected—Ross, who
had political ambitions, unloaded ITG for $99 million to Platinum Equity, a
private equity firm that specializes in leveraged buyouts. Weeks later, Trump
was elected with the promise that he would save American jobs, as part of
his Make America Great Again pledge, and he named Ross as commerce
secretary. One year later, Platinum Equity stunned the apparel industry by
announcing it was giving up on White Oak and laying off the last 208
employees. Ross made no statement.

ITG said it would keep its corporate headquarters in Greensboro and,
from there, oversee the company's textile factories in the US, China, Central
America, and Vietnam. White Oak had managed to survive offshoring, only

to be put down just as reshoring had resuscitated North Carolina's textile industry. "The White Oak facility was the last premium denim producer in the United States," Bellos said six months after its closure. "There is not a domestic mill that does selvedge right now." This left her few choices: she would still sell her indigo to Cone's plant in Mexico. And she was looking at Japan and Europe.

In August 2018, I circled back around to her to see how she was getting on post–White Oak. Better than I imagined, it turned out: Patagonia had just launched an organic denim collection dyed with Stony Creek Colors indigo.

But she had bigger news to impart: a new denim mill was opening in Vidalia, Louisiana. "They are going to fill the void left in American denim with the closing of White Oak, as well as bring something new," she said. And they'd be sourcing her indigo.

A couple of weeks earlier, the Vidalia Denim Company had purchased a nine-hundred-thousand-square-foot former Fruit of the Loom factory for $12 million from the city of Vidalia, a town of 4,000 across the Mississippi from Natchez. The project is the baby of Dan Feibus, a Scranton native in his early fifties who served as chief executive of the cotton mill Zagis USA for seven years. Like English Fine Cottons, Vidalia Denim Mills would be outfitted with state-of-the-art, computerized equipment, bringing three hundred jobs to the long economically depressed region. It would source sustainable (but not organic) cotton from Mississippi; the farmers, identical twin brothers, are investors in the project. "We can tell you the bale, gin lot—everything," Feibus told me. "The transparency is critical today."

For finishing, Vidalia hired denim wash expert Dale Jinjoe, formerly of Gap and Ralph Lauren. When I spoke to Feibus, in the fall of 2018, Jinjoe was busy building a full-scale laundry on-site, using Jeanologia-like technology that would reduce water and chemical use eightfold. Feibus believed it would be the most energy-efficient denim mill in the United States.

The plan was to put out about twelve to fifteen million yards a year—"a drop in the bucket in the denim business," Feibus said. At first, he said, they'd specialize in standard denim; Wrangler was one of its first customers. Later, it will add selvedge. ("I'd *love* to go see what they are doing in Japan," he confessed.) Like all the other rightshoring companies I spoke to, Vidalia's model "is to be extremely flexible and responsive on a quick-turn basis," he said. The factory would be up and running by mid-2019.

"It's not a boutique or feel-better-for-buying-American mill," he insisted. "We will use considerably less labor than a traditional American mill, and a ton less energy and water. This has really sound economic footing, regardless of what's happening with trade policy. Cone could not succeed with the waste of labor and materials. Most US production is a relic of a bygone time. Our view is to make extremely good yarn, and you can't do it the old way. You can't run a nimble factory the old way. It has to make sense on a big economic level, and ours does."

# part three

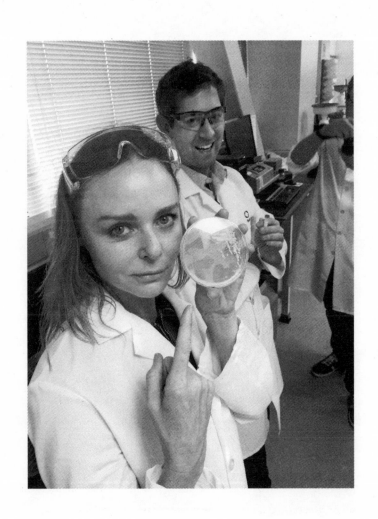

# SEVEN

# We Can

# Work It Out

THE RAIN WAS COMING down sideways. Cloaked figures in impossible shoes stepped out of a mob of black sedans on the Place de l'Opéra and, under waiting umbrellas, minced up the baroque Palais Garnier theater's timeworn stone steps. Editors. Retailers. Influencers. The tribe that decides, season after season, year after year, what's in and what's out.

On that sullen March morn in Paris, their object was British designer Stella McCartney's Fall-Winter 2017-2018 women's wear show. They descended the broad marble staircase to the opera house's lower level, chattering and air-kissing along the way, and settled on white benches encircling a small rotunda. At 9:45, the lights dimmed, and for a moment, the crowd hushed.

"Don't you fuck with my energy!" rapper Princess Nokia erupted over the speakers. Kliegs blinded. Nikons flashed.

McCartney's models strutted by in houndstooth mini-trenches, clingy caramel wool knit jumpsuits, charcoal flannel pantsuits with crisp white cotton shirts, leather biker tuxedos, and fluttering viscose sheaths muraled with rearing mustangs and nebulous blue skies. On their feet were suede princess pumps and flats. In their hands, buttery leather totes.

What the archly discerning audience couldn't catch as the models circled were the sourcing details: the wool came from a sustainable sheep farm in New Zealand, the viscose was made of cellulose from Swedish timber certified by the Forest Stewardship Council, the cotton was a heritage variety farmed organically in Egypt, the leather and suede was actually polyester and polyurethane. Scores of apparel companies present their new collections during Paris fashion week. But only McCartney bills herself as a "conscious designer." And her unquestionable commitment to those principles at the highest level of fashion has, over time, had an outsized impact on the industry.

A lifelong vegetarian and ardent supporter of People for the Ethical Treatment of Animals (PETA), McCartney has always conceived and produced clothes and accessories that are "animal-free," meaning no leather, no fur. Her supply chains are transparent and traceable. Her stores are built with recyclable materials and many are powered ecologically. Since 2013, McCartney has produced an Environmental Profit and Loss report (EP&L), which measures her supply chain's impact "from farms to finished products," as she explained it. In a business that is always in search of The New, McCartney believes being responsible is "the most modern thing that you can do."

That was not the case when she started designing in the mid-1990s. Being green has long been associated with the brown-clad, crunchy-granola set—the sort who generally eschew It Bags and Statement Looks. "I was ridiculed," McCartney told me over mint tea and cold-press apple juice at a boutique hotel near her home in Notting Hill, London, a week after the opera show. "There was anger; it was a confrontational argument."

But as sustainability and worker rights have become more mainstream, so has the public's demand for more consciously designed and made fashion. Millennial and Generation Z shoppers rank social and environmental responsibility as two of the top five factors they consider before buying a product. And according to a Nielsen global survey in 2015, 66 percent of respondents said they were willing to pay more for "products and services from companies that are committed to positive social change and environmental impact."

"Millennials want their brands to behave responsibly," says Elisa Niemtzow, director of consumer sectors for Business for Social Responsibility (BSR), the world's largest nonprofit professional network dedicated to sustainability. "They expect more from their brands from an environmental and social perspective."

"The other night I was out, and all these young girls came over to me and said, 'Thank you so much for making conscious fashion,'" McCartney told me. "That's a new thing for me."

McCartney is the ideal standard-bearer. As the daughter of former Beatle Paul McCartney, one of the most famous hippies ever, her devotion to righting social and environmental wrongs is not only genuine, it's in her blood. The second-born child of Sir Paul and his American wife, Linda Eastman, a photographer who died in 1998 from breast cancer, McCartney and her three siblings were raised on an organic farm in Sussex, England. The household was famously pro–animal rights and vegetarian—McCartney's mother penned best-selling cookbooks and launched a prepared food line that remains successful today. (Continuing the tradition, McCartney, too, is rearing her four children, with her husband, Alasdhair Willis, as antileather, antifur vegetarians.)

McCartney was a self-described tomboy, who rode her pony through the English countryside and played in streams. But she also was surrounded by fashion—her father was the most dapper of the Beatles, and her mother cultivated a cool rock-wife aesthetic—and Stella would while away the hours drawing pictures of clothes. When she was a tween, she ran up a jacket out of faux suede—the first garment she designed and produced on her own, and a portent, you could say. At fifteen, she interned in the studio of French designer Christian Lacroix in Paris as he prepared his debut haute couture collection for his new namesake brand in 1987. She later did stints at London-based designer Betty Jackson and British *Vogue*.

In 1992, she enrolled as an undergraduate in fashion design at Central Saint Martins College of Art and Design in London, the alma mater of John Galliano and Alexander McQueen. Finding the curriculum too theoretical, she apprenticed at Edward Sexton, her father's Savile Row tailor—a training still apparent in her work; Stella McCartney's suits are some of the sharpest cut in fashion.

Shortly after her degree show in 1995—which pals Yasmin Le Bon, Naomi Campbell, and Kate Moss walked for free—McCartney started a small company in London and became known for her romantic slip dresses made from antique silk and lace she found in the Portobello Road flea market.

"So, basically, your first collections were sustainable," I said.

"Exactly," she responded.

About two years later, McCartney was approached by the owners of the French luxury ready-to-wear brand Chloé in Paris to replace Karl Lagerfeld, who was stepping down as its designer. (He would maintain his jobs at Chanel and Fendi.) She knew the house well: "My mum wore Chloé in the '70s, so I had that floating around the house," she told me at the time.

During her initial meeting with Chloé executives, McCartney put forth her key design requirements: no leather, no fur. Ever.

"You take or leave it," she said. "It's not an option."

She looked at me as she recalled the moment, her aquamarine eyes deadly serious. "Stella Steel," as she has been called, is without question a woman of conviction.

Though there was resistance, in the end, they took it.

When the deal was signed and announced, in April 1997, twenty-five-year-old McCartney squealed excitedly to a reporter: "Wow! I got Karl Lagerfeld's job!" A lot of other fashion folks thought the same, albeit skeptically, and grumbled that she landed it because of who her father is rather than for her talent. Lagerfeld snipped in the press: "I think they should have taken a big name. They did—but in music, not fashion."

With her first show, dedicated to her then-ailing mother and held at the same Paris opera house in October 1997, McCartney disproved her doubters. Suzy Menkes, then the fashion critic for the *International Herald Tribune*, wrote: "Although she was following in the giant footprints of Karl Lagerfeld in Chloé's glory years, McCartney wisely sent out a simple, unpretentious show literally filled with little nothings: dresses as light as a scarf; wispy printed blouses with floaty flower-child sleeves; slithery negligee dresses, always with the dressmaking details . . . It was a fine first effort." Dawn Mello, then-president of Bergdorf Goodman, called it "absolute perfection."

Before long, McCartney was dressing such style influencers as Gwyneth Paltrow, Kate Hudson, Nicole Kidman, and Madonna, who wore a sexy pair of low-slung Chloé tuxedo pants with a sequin waistband in her "Ray of Light" video. Chloé was commanding more retail space in department stores, and sales climbed.

"What Stella did was to surprise everybody, by very, very quickly developing her own style," American *Vogue* editor in chief Anna Wintour said. "It's very much the way she dresses herself, and you can feel her in all the collections she does. We have so few women designers who are really important in the field of fashion, and it's great to have someone like Stella joining the ranks. She has made a lot of very young, very attractive girls want to buy those clothes."

But her no leather–no fur policy drew fire. Critics charged that faux hides, many of which are petroleum based, were more damaging to the earth than the real stuff.

Bull, said McCartney. "Livestock production is one of the major causes of . . . global warming, land degradation, air and water pollution, and loss of biodiversity," she shot back, with more than fifty million animals farmed and slaughtered each year just to make handbags and shoes. Conventional leather tanning employs heavy metals such as chromium, which results in waste that is toxic to humans. "Tanneries are listed as top polluters on the Environmental Protection Agency's 'Superfund'" list, a federal program to underwrite the clean-up of contaminated industrial sites, she further reported. Yet about 90 percent of all leather is chrome tanned.

"Killing animals is the most destructive thing you can do in the fashion industry," she told me. "The tanneries, the chemicals, the deforestation, the use of landmass and grain and water, the cruelty—it's a nonstarter. The minute you're *not* killing an animal to make a shoe or a bag you are ahead of the game."

McCartney quit Chloé in 2001 and launched her own label, based in London. The luxury conglomerate Gucci Group (now known as the more-friendly-sounding Kering) held a 50 percent stake; she held the other half. (In March 2019, she completed a buyback of the Kering half and now is full owner of the brand.)

With that original deal, she was swiftly accused of getting into bed with the enemy: Gucci, at its core, is a leather goods company. But to her mind, she was "infiltrating from within." Not only was McCartney planning to hold fast to her conscious fashion ethic at her own brand; she wanted to sway the group's other brands, such as Yves Saint Laurent and Alexander McQueen, to take it up, too.

To be sure, her no-leather stance led to a great deal of head scratching in the Gucci Group executive suite. After all, logo-heavy leather goods, such as handbags, wallets, and key fobs, are the cash cow of the luxury industry: they are easy to sell, instantly convey status, and boast retail markups of twenty to twenty-five times production costs. "It was like: 'How are we going to do this? Oh my God. We have to weigh the loss of leather sales,'" she remembered, during our chat in Notting Hill. "I was told: 'You're never going to have a healthy accessories business with nonleather.' I've proven that wrong."

One of the great advantages of being a female designer (and a celebrity one, at that) in fashion is you can serve as a walking advertisement for your work. With her natural beauty (she regularly will forgo makeup) and her de facto fame, which she takes coolly in stride, McCartney is most effective in this regard.

On the day we met, she was wearing a white organic cotton twill shirt that set off her sweep of russet-blond hair and her freckles, khaki men's pants "from about five seasons ago" that hung on her lithe yoga body like a boyfriend's pajama bottoms, and butterscotch faux leather sandals on her pretty and impeccably pedicured feet. "I'm a mash-up of seasons, menswear and women's wear," she said, laughing as she took in her outfit. It was all comfortably stylish and sustainable—true modern fashion.

She then showed me her handbag: a honey-hued miniversion of her best-selling Falabella. It was made of synthetic leather and lined with recycled fake suede. As I turned it over in my hands and ran my fingertips across the grain, she asked, "Really, does anyone know it's not leather?"

*Not I*, I thought.

In 2006, McCartney's company turned a profit—five years after its founding and a year ahead of schedule. A "significant" portion of those sales was from accessories, a McCartney spokesman told me; one published report estimated that they make up one-third of her turnover. (McCartney doesn't reveal such numbers.)

Having proven that going skin-free is a viable business model, she decided to see what other environmentally noxious materials could be eliminated from her line. She hit upon one: polyvinyl chloride, known as PVC.

PVC is one of the most pervasive plastics today. Cling wrap, drinking straws, credit cards, strollers, toys, artificial Christmas trees, Scotch tape, and plumbing pipes are all made of it. In fashion, it is used for transparent shoe heels, vinyl raincoats, synthetic patent leather, and the flexible tubing inside handbag handles. But it is a known carcinogen, and when it biodegrades, it releases poisons into the soil and water table. In 2010, McCartney banned all use of PVC at her company.

"Taking PVC out was a huge thing for us," she said as she poured a second cup of mint tea. "I'd say: 'Let's do a clear heel! Let's do a Perspex trench!' PVC, PVC, PVC. 'Let's do sequins!' There are two sequins in the world without PVC. There are millions of *gorgeous* sequins, but they have PVC."

By 2016, all Kering brands had stopped using PVC. For McCartney, it was a big win.

Over the years, McCartney has engaged in several independent or joint projects with more affordable apparel companies and advanced her conscious approach to fashion to the mass market—more infiltrating from within. When she designed a capsule collection for H&M in 2005, she insisted on using organic, rather than conventional, cotton. For her long-running collaboration with Adidas, she has stipulated no PVC. In 2017, she called on Parley for the Oceans, a marine protection organization that turns plastic ocean waste into thread and yarn, to help make the Adidas by Stella McCartney Parley UltraBOOST X sports shoe.

"Affordability has always been at the forefront of the way I think. At Stella McCartney, we have a lot of entry price points that, for luxury fashion, are not extortionate . . . And if you can't afford it the first time around, get it in the sale. Get it in the sale of the sale of the sale. Get it secondhand . . . Instead of buying five hundred things for X amount"—i.e., fast fashion—"buy fewer things that will last longer"—i.e., better-made, if costlier, fashion. "You will have more value attached to them, you will have more pride attached to them, you will have a better-quality product, and it will serve you well for a long period of time."

NONE OF THESE initiatives would've flown were it not for McCartney's sustainability and ethical trade chief, Claire Bergkamp. A brainy brunette in her thirties, who hails from Helena, Montana, Bergkamp is the one who finds sheep herders in New Zealand and organic cotton farmers in Egypt, liaises with pro-environment NGOs and associations, crunches the numbers for the EP&L she helped craft, and figures out how to bring impact percentages down, and down further. "I can't even really see life without Claire," McCartney said.

Like McCartney, Bergkamp is a lifelong fashion follower. She subscribed to *Vogue* throughout her youth ("I can tell you, in Montana, I was a freak!") and shopped in Idaho malls when on soccer team trips (there were no decent malls in Helena). She knew early on, however, that her interest lay more on the sociological side of dress—"What it means to wear something," she told me—than fervent consumption or preening.

After graduating from Emerson College in Boston in 2007 with a BA in costume design, and spending four years in Los Angeles working in film and television, she enrolled in London College of Fashion's master's program, with a focus on sustainability. Shortly after her degree, she landed at McCartney. Her mandate: make the company more efficiently green. "Everyone had been to our factories," Bergkamp told me, "but no one had looked at them through the lens of social and environmental impact."

We were sitting in a small conference room in McCartney's airy headquarters, situated in a 1920s brick edifice in the west London neighborhood of Shepherd's Bush. As with most luxury fashion offices, the walls were white, the furnishings were modern, and the rooms were awash in natural light—even on a dark winter morn.

One of Bergkamp's tasks at McCartney was to set up the environmental profit and loss system that Kering developed with PricewaterhouseCoopers. The EP&L analyzes six major categories—greenhouse gas emissions, air pollution, water pollution, water consumption, waste, and land use—and places a monetary value on environmental changes caused in each by the company's practices. Negatives on the balance sheet come from sourcing raw materials in a traditional manner—taking, without giving back. The plus column records actions in the "mitigation hierarchy." In order of value: avoid impact; minimize impact; restoration; offset. For example, "you can offset river impacts by cleaning up rivers," Bergkamp explained. You can "restore degraded ecosystems—like grazing land that's been wiped out—and reverse desertification by changing grazing practices." Such exercises are assigned positive monetary values and are calculated on the balance sheet.

The EP&L "allows us to set new, ambitious targets and gives us an unprecedented level of visibility and traceability into our supply chain," McCartney stated. All Kering companies now have EP&Ls.

Once the EP&L was up and running, Bergkamp could see what changes would have the greatest positive impact. They adopted the use of "Eco Alter Nappa," a new polyester-polyurethane substitute for Nappa leather sealed with a coating containing vegetable oil made from the remnants of cereal production; organic cotton for almost all denim and jersey; and ECONYL, a regenerated nylon, for handbag linings.

In 2014, McCartney publicly committed to halt sourcing rayon from suppliers who logged in ancient forests—a move that rattled the rayon industry: three-fourths of all rayon is produced by ten suppliers. To put the supply-chain change in place, McCartney partnered with Canopy, a Vancouver-based forest protection NGO. Canopy verified that all of McCartney's cellulose-based textiles made of wood pulp, such as viscose, rayon, and modal, came from sources that were free from ancient and endangered forests. Within two years of the McCartney-Canopy initiative, nine of those suppliers announced they would stop logging in rainforests.

McCartney's EP&L showed that virgin cashmere—the fiber combed off goats in Mongolia and spun into yarn—had the highest environmental impact of *all* her raw materials: roughly one hundred times that of wool. It takes four goats to obtain enough virgin fiber to make one cashmere pullover, versus one sheep to make five wool sweaters.

For centuries, cashmere has been considered a luxury product: not only was it precious to touch, it was rare. But in the 1990s, two things happened to change that point of view: Mongolia transitioned from a strictly controlled, centrally planned Communist economy to one that was market based and freer, and fast fashion—which was going global and growing exponentially—began hawking inexpensive, low-quality cashmere sweaters. V-necks. Turtlenecks. Cardigans. Pullovers. All for less than $70 apiece. These sweaters weren't as lush as the luxury versions, and they certainly didn't last as long—your elbow would poke through the knit after a season's wear. But they were so affordable, who cared? You could always buy another one. And consumers did. Copiously.

Herders responded by quadrupling the number of goats from five million in the 1990s to twenty-one million today—all on the same amount of acreage. By 2017, 70 percent of Mongolia's grasslands were degraded and desertification was setting in. "When cashmere became cheap, and fast fashion embraced it, a whole ecosystem was destroyed," Bergkamp said. If fashion's demand for cashmere keeps apace, Mongolia will have forty-four million goats by 2025.

Because of this, in 2016, Stella McCartney became one of the first major fashion companies to completely switch from virgin (or new) cashmere to "regenerated" (or "reclaimed") cashmere made from postmanufacturing waste, such as cuttings gathered from the factory floor. Regenerated cashmere is 92 percent less damaging to the environment than virgin cashmere.

The change in McCartney's EP&L was remarkable: though cashmere only made up 0.13 percent of her overall raw material usage in 2015, it accounted for 25 percent of the company's total environmental impact; after adopting the use of regenerated cashmere in 2016, the impact dropped to 2 percent.

For cotton, McCartney is just as exigent: she aims to source only organic for denim and jersey by 2025. "Luxury fashion should use organic cotton—I don't think there is any excuse not to," Bergkamp told me. "We aren't producing on the same scale as H&M. We can use organic cotton easily."

To bolster production, McCartney is participating in Cottonforlife, a five-year organic-cotton farming project in Egypt launched in 2015 by her Italy-based supplier Filmar, the Egyptian government, and Alexbank, one of Egypt's largest financial institutions. Cottonforlife is helping farmers on the Nile Delta resuscitate Giza 45 and its sibling, Giza 87, two covetable and valuable extralong staple varieties, first cultivated during the reign of Ali

Pasha in the early nineteenth century. Both are grown on fertile fields around Damietta, not far from the Mediterranean coast, without pesticides and with less water, and their transformation into cloth is monitored and certified by independent authorities. Filmar introduced the yarn, called Nilo, in January 2016, and it is available in thirty-six colors. McCartney uses Nilo for her knitwear. (Its quality is too fine for denim.)

How does this all shake out in her design?

Take that March 2017 fashion show at the Paris opera house.

The flannel suits were made of wool from a sparsely populated, open-range sheep ranch on New Zealand's South Island. "Happy sheep produce better quality wool," Bergkamp quipped. The sheared wool was shipped to Zhangjiagang, China, to be cleaned sustainably, then on to Italy, where it was spun, and woven in a factory certified by Social Accountability International, an NGO dedicated to improving workplace conditions.

The viscose for the dresses began as responsibly forested trees pulped primarily with recycled water by the biorefinery Domsjö Fabriker, in Örnsköldsvik, Sweden. From there, the pulp was transformed into filament using a chemistry-safe system at ENKA, an OEKO-TEX–certified factory in Obernburg, Germany. (OEKO-TEX is an independent testing and certification system for textiles all along the supply chain.) The filaments were then spun and woven into cloth at a mill in Como, Italy. Granted, McCartney's supply chain is still somewhat fractured, but sometimes the cleanest, fairest solutions are not all next door—at least not yet.

The clothes were cut and sewed in a Hungarian factory where workers earn a living wage in clean, safe conditions, then shipped to McCartney's fifty-plus stores worldwide.

As technology has advanced, so has the sustainability of McCartney's boutiques. Since 2005, her UK stores have been run on Ecotricity wind power, and as of 2015, all of her boutiques worldwide have LED lighting. The Dallas shop, opened in 2013, has solar panels. The Costa Mesa outpost has skylights, reducing the reliance on electric lighting. In London, at 23 Old

Bond Street in Mayfair, shelving is made of wood reclaimed from the Venetian lagoon, the wallpaper is a papier-mâché of office printouts, and the air is run through an Airlabs filtration system, removing 95 percent of pollutants. "It's the cleanest air in London!" McCartney exclaimed with pride during the store's inaugural fete in June 2018.

"We take conscious design throughout the entire company," she told me during our first meeting in Notting Hill. "Every single piece of paper, every single bag, is all recycled and recyclable." She handed me one of her shopping bags: it read on the bottom, "Made of recycled materials."

"My invitations are always biodegradable," she continued. For her March 2018 show, each included a pair of knee socks made of yarn upcycled from discarded garments and arrived in a transparent bioplastic envelope that bore the message: I AM 100% COMPOSTABLE (AND SO ARE YOU!)

"Yes, I know that the moment I create a product, any product, I'm in some way creating a footprint," she admitted. "You can't pretend that's not the case. But I always try to find a solution."


ONE WHO OFFERS SOLUTIONS is Swiss entrepreneur Nina Marenzi, founder of The Sustainable Angle, a platform that promotes the research and sourcing of sustainable textiles.

Each year, Marenzi hosts the Future Fabrics Expo in London for fashion industry players, from designers to sourcing agents. "When they walk in, their eyes light up like it's like a candy shop—all sorts of fibers and leathers and things no one has ever heard of," Marenzi told me in her Ladbroke Grove showroom. "As soon as you show designers these cool new products, they want to work with them."

We were surrounded by a dozen garment racks bearing hundreds of examples: a Naugahyde-like canvas called Pellemela that is a composite of apple juice and compote waste. And a pleasingly squishy earth-colored suede fashioned out of mushrooms. And a beautiful red "leather" that felt creamy and smelled sweet, like a fruit pie. "It's made of rhubarb!" Marenzi said with a laugh.

There was Sally Fox's colored cotton, in shades of green, brown, beige. "Imagine how much money you could save in the dyeing process if cotton was already colored," Marenzi said. And linen—a sustainability star. "It's grown on marginal land, is rain-fed most of the time, grows fast, requires hardly any pesticides, and you can blend it easily," she said.

There was cotton cloth with a waterproof coating derived from Amazonian rubber. "This helps local tribes," Marenzi explained, as she handed it to me. Its smoothness was satin-like, sensual. "They collect the latex and get some income, rather than chopping down the trees and converting the land to agriculture." There was a white crepe made of a silk-cellulose blend called Orange Fiber. It was invented by Italian students Adriana Santanocito and Enrica Arena as a way to repurpose the seven hundred thousand tons of orange juice by-product generated each year in Italy. In 2017, the Florentine luxury brand Salvatore Ferragamo used Orange Fiber for a capsule collection of shirts, dresses, trousers, and scarves.

"A majority of fashion's problems could be solved at the design process," Marenzi said. Still, she struggles to get these fabrics the attention and the orders they need to keep going. "The big firms all say they have no budgets for sustainability," she told me, somewhat exasperated. "Everyone, from the CEO to the designer, has to be on board, and a lot are not quite ready." It's a common corporate dodge that focuses on short-term profit returns rather than long-term concerns, such as the resilience of nature's bounty, without which there will be no profits.

This intransigence infuriates McCartney. "Why isn't everyone doing this? What the hell?" she railed when I conveyed Marenzi's concern to her. "Come on, guys, I have to make a bag sustainably for the same price point as you do conventionally. And I can do it. Why aren't you?"

She stopped, took a slug of her apple juice, her back stiffening.

"The fashion industry is *so* old school, it's ridiculous," she snapped.

"When are we going to wake up? When are we going to be accountable? When are we going to admit to the consumer that we are responsible for *quite* a big dent in the environment?

"How am I going to get my peers to do this?"


FASHION HAS TIPTOED into the realm of sustainability. And a good many of those forays have been nothing more than "greenwashing"—the public-relations spin that makes companies look more environmentally proactive than they actually are.

At first, the spin was pretty flagrant and, honestly, embarrassing. In 2008, a Louis Vuitton campaign featuring cultural icons, such as Keith Richards and Catherine Deneuve, posing with the brand's signature monogram luggage claimed to support Al Gore's Climate Project. When I inquired about the details with the Vuitton press office in New York, the spokeswoman told me that the brand asked its subjects to donate the fee they received for the shoot. So Keith Richards and Catherine Deneuve were supporting the Climate Project. Not Louis Vuitton.

It reminded me of what economist Robert Reich said a year earlier: when "you hear a company boast about how environmentally friendly it is, hold the applause. Under super-competitive capitalism—what I've termed 'supercapitalism'—it's naive to think corporations can or will sacrifice profits and shareholder returns in order to fight global warming. Firms that go green to improve their public relations, or cut their costs, or anticipate regulations are being smart, not virtuous."

Now, the spin is less obvious. More like a sleight of hand.

All the major brands have corporate social responsibility (CSR) officers to guide them toward cleaner, more ethical business practices. Some have their version of Kering's EP&L—at Inditex, it's called the "sustainability balance sheet," and it is published in the company's annual report.

And a good many belong to the Sustainable Apparel Coalition, a San

Francisco–based global alliance of retailers, brands, suppliers, NGOs, unions, and academics founded in 2011 by Patagonia and Walmart—the yin and yang of sustainable fashion—to assess and improve apparel supply chains. To this end, the SAC developed the Higg Index, a standardized method to measure a fashion company's environmental, social, and labor practices. Its name was inspired by the Higgs boson particle, because, SAC chief executive Jason Kibbey told me, the particle "describes the origins of the universe, and we saw this as the tool to describe the origins of your clothes." For its first ten years, the Higg Index was for industry eyes only. But in 2019, the SAC began making the index's brand and facility evaluations available to the public, and in 2020, there will be "foot-printing assessments," Kibbey said, "so consumers can see what the carbon footprint of a garment is."

Another trend is to put out ecologically minded products—like H&M's Conscious Exclusive and Zara's Join Life, capsule collections made with sustainable materials. And this is where the legerdemain comes in. Usually, such lines contribute a minute percentage of overall revenues—like in the single digits, though that's better than no digits.

Similarly, H&M has pledged to only source organic cotton, recycled cotton, or the sustainable-sounding Better Cotton by 2020. While the Better Cotton Initiative (BCI) reduces water and pesticide use and discourages child labor, it has been the target of criticism because it allows for GMO seeds, which are controversial in much of the world; it is difficult to enforce; and farmers who adopt it are not subsidized.

H&M also claims that it plans to be "climate positive" by 2040. But to be "climate positive," a company doesn't necessarily halt damaging practices; it can pay for or do good works that "offset" the damage. H&M's sustainability chief Anna Gedda told me that all these maneuvers are to "future-proof our business."

"We are going to be in this industry not for the next three years but for the next thirty years," she said. "We have to make sure we have created circumstances that allow us to do that."

Dilys Williams, head of the Centre for Sustainable Fashion at the London College of Fashion, is as dubious as Reich of such soapboxing.

"There are people shouting 'Hallelujah, we've solved the problem!'" she said when we met in her office on the university's Marylebone campus. "But you haven't even begun to solve it because you haven't tackled the heart of it: the growth model. As David Attenborough said, infinite growth on a finite planet is just bonkers. The only people saying it's possible are economists and people with a lot of money. Really what these brands are talking about is risk reduction."

"For me, sustainability is having an ecological worldview—that nature is at the heart of everything"—like McCartney, with her animal-free stance—"and the economy and the political landscape should always sit within that." Not vice versa.

In her early fifties, with a bouncy pin-curl bob, Tinker Bell eyes, and a dimpled chin, Williams is the sort of ebullient prof that students adore. "That's why I moved to London," Bergkamp told me. "Dilys was the first to teach sustainability in fashion in a meaningful way." Williams believes that education—forming the next generation of designers and assistants to think like McCartney—is the only way to bring about systemic change. After all, she said, "fashion is a means of agency."

MCCARTNEY'S MOST IMPORTANT, and enduring, act in sustainability is her godmothering of disruptive start-ups, and she does so in a variety of ways—as an investor, as a client, and sometimes simply as a promoter-cheerleader. Such is her role for the moment at Modern Meadow, an American biofabrication company that uses fermentation to grow animal-free collagen protein assembled into material inspired by leather.

Modern Meadow is as sci-fi futuristic as it sounds. And it is driven by our need.

While living in Shanghai in the late 2000s, American venture capitalist and self-described science nerd Andras Forgacs was taken aback by China's Brobdingnagian-sized demand for meat and leather goods. Already, Forgacs and his physicist father, Gabor, had founded Organovo, a biotech start-up that 3-D printed human tissue for medical use, such as reconstructive surgery, and for pharmaceutical and cosmetics testing. He wondered if "tissue engineering" could also be used to produce meat and leather-like material. He returned to the United States and cofounded Modern Meadow with his father and scientists Karoly Jakab and Françoise Marga in 2011. After discovering there were companies already well advanced in the development of slaughter-free "cultured meat" (or "clean meat"), they turned their focus to materials.

On the other side of the Atlantic, Suzanne Lee, a savvy Brit who had studied textiles at Central Saint Martins and assisted designer John Galliano early in his career, was experimenting with biofabrication—the creation of material from microorganisms—via her company, the BioCouture Research Project. When she heard of Modern Meadow, she told me it was like discovering "a whole other future that I wanted to explore." In 2014, she moved to New York to join the team, as chief creative officer.

A year later, Modern Meadow hired chief technology officer Dave Williamson from DuPont. He conducted a technology assessment of Modern Meadow's tissue engineering process—studying how thick and large they could grow the materials, how long it would take, the cost—and concluded that the process was not scalable, financially or practically. Plus, there was "the ick factor—obtaining your cells from a human or an animal," Lee said. Back then, Modern Meadow was sourcing cells from unborn calves.

The company changed direction once again, this time to an animal-free fermentation process that produces collagen protein, the main ingredient found in animal skin. If the science worked as they hoped, Lee said the scientists would be able to "create a dress in a vat of liquid."

Leather is a $100-billion-a-year business *before* it is turned into shoes, luggage, or coats. And consumer demand for leather goods is rising 5 percent annually. Lab-grown hide-like materials are seen as a viable replacement for petroleum-based faux leather and PVC-laden pleather, as well as a way to retract traditional leather's harmful supply chain; the livestock industry generates at least half of all global greenhouse gas emissions. With biofabrication there is no waste—you cultivate only what you need—and, as it is animal-free, it is vegan.

"When I first heard of it, I was kind of like, 'Hmm . . . ,'" McCartney told me, sounding somewhat skeptical. "But if you don't kill an animal, you're fine with me. You use less water, less electricity, no landmass? Bring it on. I am very excited for the day that I can have my entire store offerings made out of lab-grown leather."

On a picture-perfect summer morning in 2017, I went to see Suzanne Lee at Modern Meadow's headquarters. At the time, it was located on the eighth floor of Building A of the Brooklyn Army Terminal, a century-old former military base across the East River from Wall Street.

Like so many of the women I've met for this book, Lee is forthright, stylish, curious, and courageous. In her late forties, English pale and pole thin, with a swishy milk-white bob, she marshals teams of techies to turn futuristic fantasies into a reality that is profitable as well as beneficial to nature. "The majority of our employees—fifty-five of the sixty—are engineers," she said, as we walked down the hallway to a big open room where a sea of young people sat at rows of desks, working on laptops. "Molecular biologists, mechanical scientists, biochemists—we have twenty PhDs."

Along the corridor wall, bay windows gave a view to the lab, where black-coat-clad scientists hovered over bubbling bioreactors and trays of test tubes. "There is something intimidating about techs in white lab coats, and I wanted to get rid of that idea," Lee explained of their basic black look. "They think they are lab ninjas."

What the ninjas practice is "complex biotech," she said. "Cell engineering—designing cell organisms to produce things for us—that's advanced genetics, advanced chemistry."

But it's based on a relatively easy notion to comprehend: "how you brew beer, but instead of beer, we're making collagen, which we turn into leather." (Since my interview with Lee, Modern Meadow has stopped calling what it produces "leather," and now refers to it as "biofabricated material." In early 2019, a company spokeswoman told me it had revised its language after consulting with members of the trade. "Leather by definition is 'the tanned skin of an animal,'" she said. "So by definition, our material is not leather, because it is animal-free.")

No animal cells are needed; the DNA is written as code, like software. "We program the DNA of yeast to make collagen, so the yeast cell becomes a little collagen-producing factory," Lee said. "We're teaching an old process new tricks."

After the collagen grows, it's strained to remove the yeast, purified, and, using the biofabrication process, designed and assembled into a material that is biologically similar to leather. Like animal skins, the biofabricated material must be tanned so it doesn't rot. Chrome tanning is not sustainable, but Modern Meadow's scientists are developing a cleaner process "to suit our materials," Lee said. From modifying the cells to tanning takes approximately two weeks.

We walked back to her office so she could show me a few samples.

The first was round and thin, like a tortilla, black, and slightly rough. "We call this one 'elephant,'" she said. "It's made of pure collagen—the protein found in skin."

I caressed it with my fingertips and held it up to my nose.

"It's like real leather," I said.

She nodded.

She handed me a second tortilla-size swatch, also black, but thinner and craggy, like rhino hide. They dubbed it "Roswell," after the famous UFO mystery in 1947, because, she said, "it had an alien quality."

The third black tortilla was soft and fine, similar to Italian glove leather.

"Every month these look different," she said as she put them away. "We are learning so much at every step of the process." I spotted a sewing machine in the corner of her office. She uses it to test the material—"to make sure you can stitch and construct with it like traditional materials," she said. Eventually, Modern Meadow material will be sold commercially by the bolt for classic 2-D cut-and-sew assignments.

The company is also developing 3-D processes, such as liquid material, to fuse seams together like glue—think of neoprene wetsuits—or to pour into molds, and material grown in specific shapes. "What does the car seat of the future look like?" she asked me. "Maybe it's stitched. Maybe it's not. You can grow the Hermès Kelly bag to shape, instead of cutting and sewing it."

"We don't want to replace traditional craftsmanship," she cautioned. "We want Hermès to still make the Kelly bag by hand. But we do want to show that there is an alternative . . . [that] we can go so far beyond the use of traditional leather." Lee and her team came up with a trademarked name for their baby: Zoa, a riff on ζōē, which is the ancient Greek for "life."

A few months after my visit, in early October 2017, Modern Meadow unveiled its first prototype—a white patchwork T-shirt, the wide, irregular seams made in liquid Zoa—at *Items: Is Fashion Modern?*, an exhibition at New York's Museum of Modern Art.

Concurrently, the company hosted a two-week-long pop-up in a storefront on Crosby Street in SoHo, to explain to anyone who dropped in what Modern Meadow and biofabrication are all about. On the wall, there was a mural-like timeline, from the Stone Age to synthetics; scientists believe the first pair of leather shoes were fashioned and worn in 170,000 BC. In the back, a minute-long video about biofabrication played on a loop. It asked viewers: "When was the last time you wore a revolution?"

Throughout the space, Zoa prototypes—conceived by Lee, senior materials designer Amy Congdon, and their team at Modern Meadow—were exhibited in waist-high display cases. Such as a black collage of cotton, mesh, and Zoa in the form of a T-shirt. A swatch of double-faced cotton with black-and-white liquid Zoa poured together in a swirl. A filmy silk square of cloth spray-painted with liquid Zoa. A spacer mesh seamed with liquid Zoa.

With the help of substantial fund-raising—$53.5 million in private investment, including from Horizons Ventures and Iconiq Capital, and $33.9 million in grants and tax credits—Modern Meadow has since left its Brooklyn Army Terminal digs. The Design and Applied Research Studio has moved to New Lab in the Brooklyn Navy Yard. The biotechnology, fermentation, and material science laboratories—and the ninjas—are now in an immense Brutalist building in Nutley, New Jersey, that formerly housed Hoffmann-La Roche pharmaceuticals.

In 2019, Modern Meadow planned to introduce its first product made of Zoa, designed hand in hand with a luxury house (not McCartney). It was also working with a performance-sport brand. Modern Meadow joined forces with the companies' research and development divisions to build the items together; this is how Modern Meadow wants to work with all its partners.

Forgacs and Lee are particularly excited about Modern Meadow's possible applications in luxury fashion—so much so, they hired a trio of industry veterans as consultants: Anna Bakst, former group president of accessories and footwear for Michael Kors (she also joined the board); François Kress, former president and CEO of Prada USA; and Mimma Viglezio, former global corporate communications director for Gucci Group (now Kering). For Viglezio, Modern Meadow's pitch is crystal clear.

"People will ask, 'So why am I buying Prada instead of Gucci when it looks exactly the same and is the same price?'" she said.

Because, if the Prada is made of Zoa, it "is better for the world."

Up in the eaves of the MoMA exhibition, another tech wonder was on display: a billowing goldenrod knit cocktail dress designed by Stella McCartney and made of lab-grown "spider silk" produced by the Bay Area–based biotech firm Bolt Threads.

With her animal-free stance, McCartney has long struggled to justify her use of silk—to herself as well as to her followers. Raw silk is spun into cocoons by *Bombyx mori* silkworms, and to recover the thread undamaged, cocoons are submerged in boiling water, killing the larva. Over the years, McCartney has tried alternatives, such as silk collected after moths have hatched, but the quality is considerably inferior. "Finding Bolt has been a life-changing and career-changing moment for me," she said.

Bolt began as doctorate theses for University of California, San Francisco bioengineering students Dan Widmaier and Ethan Mirsky. In the midaughts, they met in the synthetic biology laboratory while experimenting with spider silk. A third PhD candidate across the Bay, University of California, Berkeley bioengineering student David Breslauer, was also trying to re-create silk in a lab and reached out to Widmaier and Mirsky for protein. The threesome teamed up and, in 2009, launched Refactored Materials in a San Francisco incubator to turn their experiments into a proper business.

Spiders extrude seven different sorts of silk, from capture-spiral, used to nab prey, to egg cocoon, to protect egg sacs. To understand how they wove their webs, Widmaier, Mirksy, and Breslauer filled an office with golden silk orb weavers (*Nephila*) and hula hoops and watched the creatures do their thing. The scientists deduced that dragline—the strongest silk, which arachnids use to drop from a fixed place and dangle—would work best as the inspiration for their first manmade silk thread. Funding arrived from Steve Vassallo of the Silicon Valley–based Foundation Capital.

The science behind Bolt's spider silk is similar to Modern Meadow's leather-like material: geneticists and microbiologists at an independent laboratory in the Bay Area design the DNA sequence to create the silk protein. Test tubes containing the DNA are FedExed to Bolt in Emeryville. "Life comes in a box delivered to your door," Jamie Bainbridge, Bolt's vice president of product development, mused to me on my visit to Bolt Threads in 2017.

The DNA lands in Bolt's biology lab, where it is introduced to yeast—like at Modern Meadow, yeast serves as the DNA's host. At Modern Meadow, the DNA is programmed to make collagen; at Bolt, to produce silk. The yeast is shocked with heat and cold, which allows it to accept the new DNA. "There are days it smells like a bakery," Bainbridge said as we walked through the lab.

Behind us were freezers set at minus-80 degrees Celsius—one was named Antares, another Betelgeuse—where the scientists store promising strains for future use. "Our library," Bainbridge said.

We crossed the corridor to the fermentation room. There, the yeast is transferred to fermenters—stainless-steel contraptions that look like stills, with rubber tubes and glass containers that percolate. "Standard equipment in beer brewing and insulin," she said. These were the two-liter versions. From there, the brew is put into 100-liter fermenters. The DNA-infused yeast is fed dextrose (sugar) and water and taken through a fermentation cycle. The silk grows. The cell broth is filtered and centrifuged. What remains is pure silk. It looks like shiny, champagne-hue taffy. Very Willy Wonka.

The taffy is run through a spray dryer—"like you use to turn liquid milk into powdered milk," she explained—and stored. She pointed to a plastic sack the size of a soccer ball, filled with about two kilograms of white powder, sitting on a shelf. "That's a bag of silk," she said. One hundred twenty grams of the powder creates one square meter of cloth; that two-kilo bag contained about seventeen meters' worth.

When it is time to spin some, the powder is mixed with a proprietary solution to turn it into a soluble dope, "like molasses," Bainbridge said. In a process called "wet spinning," the dope is loaded into a pump, extruded through a spinneret—"the exact name of the back end of a spider, where the web silk is pulled out," she said—and plunged into an aqueous bath to coagulate. It is colored with textile dyes and spun on a series of godets—like spindles—each one whirring faster and pulling tighter to crystallize the fiber and build the yarn's structure. Then it is wound onto a spool. Finished.

She showed me several. By touch and sight, I could not tell it wasn't natural silk.

Once the team got to that point in development, it was time to scale up. They hired Nike's former US women's brand director, Sue Levin, as chief commercial officer; she immediately changed the name from Refactored Materials to the stronger-sounding Bolt Threads. When they had a bona fide lab-made yarn, which they christened Microsilk, they called on local master weaver Lillian Whipple to turn it into fabric. She produced a few swatches, and with each iteration, Bolt's scientists adjusted and refined the extruding and spinning process until they had a fine-looking piece of cloth. But when they washed it, it shrank by roughly 40 percent.

In early 2015, they moved from the incubator to the five-story building in Emeryville, a waterfront town next to the Bay Bridge, between West Oakland and Berkeley, and brought on Bainbridge. A Seattle native in her fifties, with shoulder-length silver hair and blue-rimmed prof specs, she, too, had worked for Nike—as head of the advanced R&D group for apparel. To solve the shrinkage issue, she suggested blending the spider silk at the polymer stage with a cellulose called Lyocell, "the cleanest of the rayon processes," and "biodegradable at end of life," she said. It worked.

A few months after landing in Emeryville, Bolt received an email from Claire Bergkamp in London. "I want to know more about what you are doing," she wrote. "It's aligned with what we are doing." After a few phone conversations, she flew out to the West Coast to see them. "I was blown away," she recalled. "It wasn't innovation for innovation's sake."

That summer the Bolt team went to London to meet McCartney, and the two sides signed a joint development agreement. At the end of 2016, McCartney paid a visit herself. "When we go to Silicon Valley, I feel we are at home," she told me. "We work in the way they do, but in fashion. It's how we look at life."

About the same time, Paola Antonelli, senior curator of the Department of Architecture & Design at the Museum of Modern Art contacted Bolt

Threads to say she wanted to include the company in her upcoming exhibit, *Items: Is Fashion Modern?*—the same show that Zoa was in. Bainbridge proposed they join forces with McCartney, and Antonelli agreed. So did McCartney.

Bainbridge sent McCartney several existing colors of Microsilk; she fell hard for the luminous goldenrod. It had a warm glow to it, like a summer sunset. She designed a halter-top dress gathered at the neckline by a wide drawstring, the sashes dropping down the back, and had her knitter in Italy make a prototype in woven linen as a guide. The gold Microsilk was spun in Bolt's Emeryville labs, then, following McCartney's specs and with the prototype in hand, Oakland-based custom knitwear designer Myrrhia Resneck produced the dress on a traditional double-needle flatbed machine. It was sent to New York and suspended in the exhibit.

What I saw in Emeryville was small-batch research and development. Big orders are contracted out to industrial-sized fermentation facilities in the Midwest, where the fermenters—originally constructed in the 1990s for biofuels—stand five stories tall and hold a hundred thousand liters of yeast-sugar-protein soup. The finished silk is sent back to Emeryville, where it is woven in-house—either on machines in the building I visited or at the company's new weaving facility down the street. Resneck is now a knitwear engineer there.

In the spring of 2018, McCartney's second Bolt Threads–sourced outfit—a blouse and a pair of pants made of a Microsilk-cellulose blend in chocolate-mousse brown—went on display in *Fashioned from Nature*, a sustainable materials exhibit at the Victoria and Albert Museum in London. Next to it was one of McCartney's Falabella handbags, black with a metal chain trim, made of Mylo, Bolt's newest material. Mylo is a faux leather made of mycelium, the underground root structure of wild mushrooms. Bolt developed it in partnership with upstate New York biomaterials start-up Ecovative.

The mycelium grows in a tray until it is about three inches thick and spongy as marshmallow. It is then compressed, dyed, and, because it is a living organism, tanned (using a sustainably certified process) "to render it 'imputrescible'—meaning it's not going to rot," Bainbridge explained. In the fall of 2018, funded by a Kickstarter campaign, Portland-based bag designer Chester Wallace produced a handsome tote called the Mylo Driver Bag. Offered in a limited edition of one hundred, for $400 apiece, it sold out quickly. McCartney expected to have items made of Microsilk and Mylo in her stores by the end of 2019.

By mid-2018, Bolt Threads had raised $213 million—investors include Peter Thiel's Founders Fund, former Google CEO Eric Schmidt's Innovation Endeavors, and Edinburgh-based asset manager Baillie Gifford—and was valued at more than $700 million. Bainbridge told me the company was working on a third protein-derived material, to be introduced in late 2019. "We can use any functional protein of an animal in this process," she explained. "The beak of a squid to catch its prey is one of the strongest known to man. The elastomeric between the wings of a fly to allow its wings to beat faster—like a rubber hinge on a door. The colored patterns in coral—this would allow us to introduce color without using dye."

When I last spoke with Bainbridge, in October 2018, she was in Florence, Italy, with Claire Bergkamp, meeting with manufacturers to see how Microsilk fiber ran on industrial spinning and weaving machines—the first step in building a supply chain. I asked Bainbridge if she planned to take Microsilk and Mylo to Première Vision.

"PV is such an old-world way of conducting business," she said. "You must have business with a mill before they will let you in. It is the antithesis of a modern way to introduce a material to the world."

As I would soon find out.

# EIGHT

# Around and Around We Go

Seven months after my first visit, I was again roaming the enormous halls of Première Vision, in September 2018. The target of my mission this time was Smart Creation, the new sustainable materials corner of the trade show—an atoll of start-ups in the sea of Big Textiles—to meet with the founders of Evrnu, a Seattle-based research and development firm that takes old cotton T-shirts and jeans and recycles them into pristine new fiber for cloth.

I finally found Evrnu along the back wall of Hall 3, a thirty-minute walk from the prime real estate where the Como mills reside. Smart Creation housed a dozen or so booths for eco-textile firms such as Frumat, the apple-waste pleathermakers, and SeaCell, which produces a seaweed-cellulose blended fiber.

Presented simply as the latest technology, these companies promise a reality that other exhibitors might prefer to keep out of sight. Namely, we can't keep making so many clothes—at our current pace, the amount is expected to increase 63 percent, to 102 million tons a year, by 2030—or discarding them at the rate that we do now. We can't keep thinking linearly: birth of a product, use of a product, death of a product—what the authors Michael Braungart and William Mc-Donough called "cradle-to-grave" in their best-selling book, *Cradle to Cradle*, and industry players describe as "take-make-waste." We must graduate to a circular—or closed-loop—system, in which products are continually recycled, reborn, reused. Nothing, ideally, should go in the trash.

Dame Ellen MacArthur, Britain's most successful offshore sailing racer, has dedicated herself to making circularity a common practice—with a keen focus on the fashion industry. In 2009, she founded the Ellen MacArthur Foundation, a charity on the Isle of Wight dedicated to researching challenges that face the global economy. She has spoken on circularity at the World Economic Forum and the Copenhagen Fashion Summit, which is considered the Davos of sustainable fashion. And her center collaborates directly with business, government, and academia to "build a framework for an economy that is restorative and regenerative by design," as her spokesman explained. The foundation offers circular design workshops and courses for everyone from students to fast-fashion executives—H&M's Anna Gedda told me she was taking a few.

At the Copenhagen summit in 2017, MacArthur's foundation launched the New Textiles Economy initiative (now known as Make Fashion Circular), a program that aims to build a circular economy for textiles, beginning with clothing. Six months later, in partnership with Stella McCartney, MacArthur released "A New Textiles Economy: Redesigning Fashion's Future," a decisive 150-page report by her foundation that laid out why it's imperative that fashion embrace circularity. Of the "tons of clothing produced each year . . . 87 percent of that gets land-filled or incinerated" and 1 percent is recycled, MacArthur told the Copenhagen conference's audience. All of this, she said, "shows a very broken system."

To tack in the right direction, she outlined a three-point plan:

One: "Build a fashion system/fashion industry whereby the inputs—the materials—are safe, nontoxic, and renewable."

Two: "Make higher-quality garments. Make reparable garments."

Three: Design apparel that, "at the end of that system, can be turned into new clothing."

With today's "innovation and creativity," she said, "why would we ever design anything that could be waste?"

Fashion industry veteran Stacy Flynn posed this last question to herself nearly a decade before MacArthur's Copenhagen speech, while on a sourcing trip to China. And it propelled her to invent Evrnu, a molecularly regenerated fiber made of 100 percent postconsumer cotton garment waste. Evrnu consumes 98 percent less water than virgin cotton; produces 80 percent less greenhouse gas emissions than polyester, viscose, or elastane (i.e., Spandex, Lycra); emits zero plastic microfibers; causes zero deforestation; and requires zero farmland.

Flynn had worked for Big Fashion for her entire career—DuPont, Target, Eddie Bauer. When she went to China for Target, she always visited top-tier factories. "You could eat off the floor of the facilities," she said. "I was like, 'China's lovely, China's pristine, we have no problems in China.'"

But when she returned in 2010 to see subcontractors in north Xiamen on behalf of a small American textile firm, she saw the other side—the hidden side—of apparel manufacturing. "At ten o'clock at night we go into this facility. There's one light source in the center of the room. There are people sewing all over the place, in the dark. There's an area where they have fabric spread out and a skull-and-crossbones on the chemistry. A guy smoking right in front of it, where it says, 'Flammable.' I'm like, 'Where am I?' I backed into a wall, and it started crumbling. It was grim—the land of the living dead. All the rules were being broken. It opened my eyes to what low price does," she told me at her PV stand. "Then I began adding up how many billions of yards of fabric I'd made up to that point in my career, and I became linked to the cause."

She returned to Seattle, enrolled in Bainbridge Graduate Institute (now Presidio Graduate School), and earned an MBA in sustainable systems. During her studies, she learned that 90 percent of all clothing is made from two fibers—polyester and cotton—and most of it, as MacArthur stated, ends up in landfills. "So the bookends were the problem: the resource extraction on the front end and the waste on the back end," she said. "The design challenge was clear: take the waste and turn it into high-quality new fiber."

She decided to take on that challenge and called her friends in fashion to see who else was game.

"Everyone said I was crazy," she told me. "Couldn't be done."

Forty-five and elfin, with a mop of soft platinum curls and an energy so positive and vibrant it could light up a Christmas tree, Flynn is the essence of American can-do. "If Stacy sets her mind to something, she'll make it happen," Claire Bergkamp told me. "You can be sure of that."

In fashion and apparel, Flynn said, "we fail to innovate on so many levels because we've been reliant on nineteenth-century equipment"—spinners, looms, sewing machines—"and the way we think about that equipment is with a twentieth-century mind-set—that resources are infinite, that cash is the only thing that matters. Everything is based on style obsolescence, with consumption being the key driver. If I can give one solution that allows brands to get their heads out of their asses to think about things in a different way, we're going to make some real progress. Because nobody *wants* to create damage."

Perhaps. But Flynn and her partner, industry vet Christopher "Christo" Stanev, also understood that they had to "find solutions that could alter [the system] without asking consumers, brands, or makers to change too much at one time."

In other words, baby steps. In 2012, they formed their first company, called Future Resource Collective. They poured in their own money and set to unlocking the science. They figured out how to purify the existing cloth and strip it of chemicals, such as aluminum from deodorant. Then they introduced a solvent that converts the cloth into liquidy cellulose—a honey-like goo that, like at Bolt, is called "dope." (They recover, purify, and reuse the solvent several times.) Also like at Bolt, they run the dope through extruders—which Flynn and Stanev designed—to turn the cellulose back into fiber. "Visualize making pasta," Flynn told me. "You start with dough, then you push the dough through a dial that makes the shape." In 2014, after filing their first provisional patent, they created the company now known as Evrnu—"like a license plate," she said.

She reached over to her display table and picked up a ball of shimmery white fluff.

"It's made of 51 percent Evrnu and 49 percent cotton," she said, as she handed to me.

I found myself compulsively stroking it.

"It's like petting a baby bunny," I said.

"I know, right?" she said, laughing.

In 2015, Flynn and Stanev had taken Evrnu on the road. At meetings, Flynn would place three beakers on the conference table. One contained shreds of her favorite college T-shirt. The second: the T-shirt jersey, dissolved. The third: a crude monofilament fiber made by extruding the cellulose through a syringe. "It was literally the easiest, lowest-cost prototype we could make to demonstrate feasibility," she recalled.

She'd tell executives: "I'm going to change your business with this technology."

Again, everyone thought she was crazy.

At Target, her former boss said: "Stace, we're a $70 billion company and you literally have three beakers and a dream. You've really got some guts."

Flynn attended a Fashion Institute of Technology conference in New York where Levi's innovation head Paul Dillinger gave a keynote address.

"And he said, 'Anyone who can figure out how to create a high-quality re-cycled cotton is going to basically be queen or king of the universe.'"

Afterward, Flynn walked up to Dillinger, introduced herself, and handed him a piece of Evrnu yarn dyed with organic indigo.

"This is made from 100 percent regenerated postconsumer garment waste," she told him.

He touched the sample.

"We need to talk," he said.

In 2016, Evrnu and Levi's unveiled the first two pairs of 501s created with regenerated denim made of a blend of organic cotton and Evrnu. The denim was hand-dyed in Seattle by Kathy Hattori of Botanical Colors. The initial prototype was hand woven by Patrice George of the Fashion Institute of Technology in New York. The wear-test versions were mechanically woven at North Carolina State—the same university where Natalie Chanin stud-ied, Sarah Bellos interned, and Sally Fox conducts her research and spins her cotton. Flynn and Bellos were going to meet the following month at a two-week accelerator in Connecticut hosted by Unreasonable Impact, a pro-gram founded by Barclays and the Unreasonable Group to support entrepre-neurs. She was excited to talk to Bellos about working together. Evrnu takes dye so efficiently it reduces the average amount needed by 20 to 40 percent, which, she said, "could make natural botanical-based systems like Sarah's affordable."

The first Evrnu Levi's were "a little industrial miracle," Dillinger told me proudly. Since then "we've gone to viable, wear-testable garments that have the same durable strength properties and wearer's experience as a conven-tional cotton. We haven't run it to market yet, because it's an R and D inten-sive process, but we are thrilled with the progress, and we are now just shy of the point where we introduce it into our broader supply chain as material."

Flynn and Stanev have raised $6 million in funding—mostly since the Levi's project. Flynn conducts the initial research and development at the company's lab in Washington State. Stanev is responsible for all research and development and technology at their laboratory in New Jersey. At Evrnu's pilot facility in South Carolina, "we extrude prototype fiber on behalf of our brand/retail partners," Flynn said. More rightshoring.

After Levi's came Stella McCartney.

Flynn met Bergkamp at a Cradle to Cradle event in 2015, and they stayed in touch, Flynn regularly sending updates on the technology's progress. In 2017, McCartney signed on as an "early adopter," supporting and testing experimental versions of the product. When I met Flynn at PV in 2018, she had in her hand her first prototype for Stella McCartney—a four-inch square of black single-weave crepe made of 51 percent Evrnu and 49 percent long-staple cotton. She planned to present it to McCartney's rep the next day. It was handsome and looked like a fine silk crepe. Originally, McCartney wanted a double-weave, but the machines that produced double-weave in the US were sold off when the domestic textile industry collapsed in the 1990s. McCartney was happy enough with the swatch to move forward and develop fabric with Evrnu in Italy, where there are still weaving machines that do double-weave.

Flynn has also signed with her old colleagues at Target and with a major athletic wear company that was still on the QT. By late 2019, Evrnu should be on the market.

The most important part of Evrnu, Flynn said, is this:

"Everything we're building we're designing to be disassembled in the future."

And on this point, she refuses to bend. One brand asked her to include Spandex—the elastic embedded in fabric to make clothing more forgiving—in the mix, and she said no.

The brand pressed. "Seventy percent of the global assortment has Spandex in it," executives argued. "We *have* to use Spandex."

Flynn still said no.

"Spandex"—a complex copolymer—"can't be recovered or broken down," she explained.

The impasse pushed Flynn and Stanev back into the lab, and they developed an Evrnu stretch fiber that can be regenerated. "We want to take full responsibility for everything we create," she told me. "We want to guarantee everything that's got an Evrnu tag on it is not only made from your own clothing but can be broken down again."

In terms of raw materials, the potential supply for a company like Evrnu is endless. There are all those millions of tons of clothes we throw away; in New York City alone, apparel and textiles make up more than 6 percent of all garbage, which equals nearly 200,000 tons a year. Rather than just fill dumps, that waste can be rounded up by companies like I:Collect—or I:Co, for short—a Germany-based global collection and sorting service for discarded apparel and shoes.

There are the clothes we give away: of the 15 percent of our clothes we donate—rather than throw in the trash—only about a fifth actually goes to those in need. Goodwill et al. shrink-wrap the remainder in bales the size of your fridge and sell them to for-profit textile recyclers, who turn the cloth into mattress stuffing, insulation, or rags.

Flynn can also use "preconsumer waste"—the scraps on the factory floor. And "deadstock"—the leftovers that brands traditionally burn or shred. When we met, she was in talks with a major sanitation company.

Once Evrnu hits its stride, she and Stanev have a plethora of circular ideas they mean to advance. "Christo and I have mapped this out for the next twenty years," she said. "Our goal is to go through the entire supply chain and create innovations that are powerful enough to scale in our lifetime. I have this written out to the year 2050. I'll be seventy-seven years old.

"Then the kids can take it," she said. "When the kids that are in the world right now come to me and say, 'What did you do when you knew?' I'm going to say, 'I changed the way I think and act, I fought like hell, and I influenced a whole bunch of other people to change the way they think and act. We didn't sit on the sidelines. Good luck.'"

WHILE FLYNN and Stanev were in New Jersey, turning cotton into dope, Cyndi Rhoades, an American expat in London, was wrestling with another, and more complicated, circularity issue: how to divorce polyester from cotton in textile blends for reuse.

Synthetics burbled forth in the 1930s, when American chemist Wallace Carothers, head of DuPont laboratories in Wilmington, Delaware, began experimenting with polymers; neoprene, artificial silk, and nylon are among fabrics born from his work. In 1941, British chemists John Rex Whinfield and James Tennant Dickson, of Manchester, built on Carothers's advances to create polyethylene terephthalate—also known as PET or PETE—the basis for polyester, Dacron, and Terylene. Synthetics went mainstream during World War II, when staples such as cotton and wool were requisitioned for the war effort and the textile industry needed alternatives for street clothing.

Like so many inventions, synthetics were considered a wonderful breakthrough and the perfect solution for a need at the time. But nearly a century on, it is clear that polyester and its cousins are far from being a cure-all. Polyester, the cheapest and most popular of fabrics, is petroleum based; nearly seventy million barrels of crude are required to make the virgin polyester used for textiles each year.

Cotton-poly blends represent about one-third of all cloth produced today. Because no one had figured out how to disentangle the two, garments made of this fabric were traditionally doomed to be dumped. And polyester is not biodegradable.

"Wouldn't it be smarter," Rhoades reasoned, "to put it back in the system?"

A decade on, she is doing that with Worn Again Technologies—a company that has developed a patented process that separates, decontaminates, and extracts polyester and cellulose (from cotton) from pure and blended fabrics and converts the polymers into virgin-quality polyester. (They also convert the cellulose, like Evrnu does.)

"Circular *is* the future," she said. "It brings fashion head-on with reality."

We were on the Eurostar with her chairman, Canadian business executive Craig Cohon, hurtling across the French countryside to Paris on a summer afternoon in 2017. Rhoades and Cohon were on their way to meet with potential investors, including Miroslava Duma, the Russian fashion editor turned digital entrepreneur.

Unlike Flynn, Rhoades was a stranger to fashion.

Born and raised in Columbus, Ohio, she moved to L.A. in the early 1990s and worked in music videos. She landed in London in 1993 for what she thought would be a three-month sabbatical and never went back. She kept producing music videos—for Oasis, Tina Turner—and documentaries for British government agencies. "Propaganda!" I teased her. "Yes, propaganda." She said with a laugh.

She read Naomi Klein's *No Logo* and David C. Korten's *When Corporations Rule the World* and "went down a road of self-teaching on global economics and what made the world tick," she told me. "I wanted to unpick how fashion is made and food is made."

To spark others to do the same, she founded Anti-Apathy, a regular salon-like gathering in an East End nightclub called The Spitz, with guest speakers, musical performances, film screenings; her inaugural edition featured the then-unknown indie band the Libertines. Each event had a theme—food, money, fashion. "We were looking at the positives—what people could do, how they could change it up, vote with our wallets, shop differently, and influence brands to behave differently."

One night in 2005, Galahad Clark, a seventh-generation member of the Clarks shoe dynasty, attended Anti-Apathy and was so affected by its message, he asked Rhoades if he could produce and sell a line of footwear to help underwrite the cause.

"Yes," she said, "as long as they're made of secondhand materials."

Together, they put out a collection of shoes that were upcycled—meaning recycled into a higher-quality product. Called Worn Again, the shoes were made from used British military jackets and prison blankets, at factories in China, and sold under Galahad Clark's independent company, Terra Plana.

After a few seasons, Rhoades took full control. "I wanted simpler products manufactured in the UK," she said. She came out with a range of bags, accessories, and windbreakers conceived by British designer Christopher Raeburn and cut from retired Virgin hot air balloons and Eurostar uniforms. Another product, designed by Benjamin Shine for Eurostar, was a train manager's bag for conductors, also fashioned out of old Eurostar uniforms. All of it was made in Great Britain.

One of her investors was Cohon. Tall and suave, he was an old-school corporate capitalist, descended from an old-school corporate capitalist: his father was the founder and chairman of McDonald's Canada and McDonald's Russia. In the early 1990s, shortly after the fall of communist rule, Cohon was charged with taking Coca-Cola to Russia, and later he started and ran Cirque du Soleil Russia.

In 2000, he saw Bill Clinton speak on the role of business in society at the World Economic Forum in Davos, Switzerland. By then, Cohon was in his thirties and Coca-Cola's deputy division president for northwest Europe. "It was the time when the antiglobalization movement was just starting—there were riots in Seattle, in Davos—and [Clinton] said we needed a kinder, gentler capitalism," Cohon remembered. "I felt he was talking right at me."

Cohon quit his job on the spot. Ever since, he's dedicated himself to bettering humanity and the environment. He's done charity work in London, Toronto, Jakarta, and Johannesburg, and partnered with BP in India to create a clean-energy stove fueled by agricultural waste and cow manure; the stoves were offered to women in rural India for $75 in microcredits. "We've sold millions of them," he said. In the late 2000s, he connected with Rhoades through a mutual acquaintance. Impressed with her upcycling projects, he invested.

While on a sourcing trip in Japan for McDonald's—"They wanted to make their uniforms out of recycled materials," Rhoades said—she visited Teijin, one of the first companies to recycle polyester. Their system, called Eco Circle, was "too closed-loop" at the time, she said; they could only recycle their own fabrics, "and it was waaaaay too expensive" for the markets she was exploring.

But it "switched a lightbulb on."

It was clear to Rhoades and her technology partner, Nick Ryan, that "second-life products"—like the anoraks made of old hot air balloons, or rags made of old T-shirts, or insulation made of old jeans—"don't solve the textiles waste problem," she said. "They still end up in a landfill."

"We thought: 'These guys are onto something: breaking down materials to their original components and putting them back together again.'"

Surely, there was a more efficient and economical way to do it.

Dr. Kate Goldsworthy, of the Centre for Circular Design at the University of the Arts London, introduced Rhoades and Ryan to Dr. Adam Walker, a Cambridge PhD who specializes in polymer chemistry.

Basically, Rhoades said, "he takes things apart."

In his lab at the Welding Institute in Cambridge, England, Dr. Walker had already dissected such complicated chemical mélanges as polystyrene, one form of which is Styrofoam.

"We want to separate polyester and cotton," Rhoades told him.

"I can do that," he responded.

And it was at that moment Worn Again "evolved from an upcycling company to a technology company in the circular economy," Cohon said. He became more involved, supporting the business side. Dr. Walker was named chief scientific officer.

The process Walker developed is relatively straightforward. Pure and cotton-poly blended fabrics are relieved of their zippers, buttons, rivets, and metals and are entered into a process that strips away the dye. Solvents are used to dissolve the polyester. The solvent is removed and recycled, and what remains is what Rhoades describes as "a hot gooey melt" called polyester tetraacrylate—or PET. It is extruded and chopped into pellets that look like shiny rabbit food. The pellets are sold to mills, where they are re-melted and spun into yarn. The cellulose recaptured from cotton can also be spun into fiber for new textiles.

In 2012, after about six months of research and testing, Cohon and Rhoades had something to show to potential customers. Like Flynn, they went to meetings with beakers: one filled with PET, the other with cellulose powder. They acknowledged that synthetics had been recycled before Worn Again, in a practice called "mechanical recycling"—but that system only handled monofibers, the quality was markedly lower, the retained PET kept the material's original color, the process could only be done once, and it could be costly. Worn Again aimed to produce PET at the same price, same quality, and same exact chemical makeup as virgin polyester. "To break the system, we had to have cost parity," Cohon told me. "Consumers have to be able to afford it."

While their original science focused on blends, it could be applied to all polyester fabrics—a sizable "feedstock," as they say in the business, since polyester is present in 60 percent of clothing. In 2016, that equaled 21 million tons—a 157 percent increase from 2000.

Worn Again's first pitch was to H&M, in Stockholm. The second was to Puma, which then was a part of Kering. "Both [H&M and Kering] put money in," Cohon said. H&M is an investor, with a stake in the company; Kering contributed to research and development.

"If we are going to tackle this issue, we have to get in bed with these guys," Rhoades said, implementing Stella McCartney's "infiltrating from within" doctrine. "Yes, there's still growth," but "we can minimize impact from the beginning of the process."

In 2018, Worn Again was the first chemical-recycling technology to receive Cradle to Cradle certification. The sustainability seal of approval came with a mid-five-figures grant. And Rhoades and Cohon met their $6.7 million fund-raising target, with investments coming from Swiss chemical engineering giant Sulzer Chemtech, Mexican garment manufacturer Himes Corporation, Mexican wholesale fabric supplier DirecTex, and Miroslava Duma's Future Tech Lab. With Sulzer Chemtech, Worn Again is conducting industrial trials and designing full-scale production, in hope of opening its first industrial demonstration plant in 2021, most likely in western Europe. Eventually, they plan to build factories near garment collection hubs.

Rhoades has conquered cotton-polyester blends and is commercializing both; Flynn, at Evrnu, has the technology for polyester and is considering commercializing it as well. "Hopefully we'll both get to both places," Flynn said. The women are friendly competitors, with great reverence for each other. ("A pioneer," Flynn calls Rhoades. "A firecracker!" Rhoades says of Flynn.) And they are perfectly positioned to dominate the market. Virgin PET and cotton were a $30 billion business in 2018; that number is projected to surge by 63 percent by 2030.

"This is not about fashion, really," Rhoades said. "It's about the future of raw materials."

PERHAPS THE BIGGEST NAME in reconstituted raw materials is ECONYL—"eee-co-kneel"—a regenerated nylon composed of used carpets, old fishing nets, and fabric scraps.

ECONYL is the circular baby of Giulio Bonazzi, chairman and chief executive of Aquafil, a nylon manufacturer founded in Arco, Italy, in 1969. Like polyester, nylon is petroleum-based. Because Aquafil is located close to Lake Garda—Italy's largest—Bonazzi told me, "we were always careful not to create too much environmental damage and we wanted to be eco-efficient." The company began recycling nylon waste in the 1990s; in 2007, it began exploring how to recycle all nylon. "Of course, being less bad is always better," Bonazzi said. "But we wanted to change to being good."

After four years of research and development, Aquafil unveiled ECO-NYL. Like Evrnu and Worn Again, Bonazzi said, "we take waste from all over the place." For fishing nets—there are some 640,000 tons of abandoned nets in our oceans—Aquafil receives calls as far away as Australia and New Zealand. "Next week, I will be in Japan organizing net collection on the northern part of the Hokkaido island," he said at the Copenhagen summit in May 2018. There is also, surprisingly, an enormous supply of used nylon carpeting—Aquafil collected 75 million pounds (34 million kilos) a year, Bonazzi said in late 2018. And once he opened his two new carpet collection-and-recycling plants in North America—one in Phoenix, the other outside of Sacramento—that would jump to 100 million pounds (45 million kilos) annually.

The nets and the carpets are "decontaminated," he said, then the nylon is stripped out and sent to the ECONYL Intake Center in Slovenia, where "we put them in a—let's call it a magic box because it's complicated"—and it is broken down into a monomer. Unlike Evrnu and Worn Again, this is done "without any solvent—only temperature, steam, and energy, all of which are 100 percent renewable," he said. The monomer is then built back up to a polymer, which is sold to spinners, weavers, tufters, and molders. "We can produce any color, or make it dyeable—whatever you want."

Surfing star Kelly Slater was one of the first in fashion to embrace ECO-NYL, for his sustainable label, Outerknown. "I knew that I wanted to certify that we were using responsible production practices, [providing] good working conditions for people, and sourcing as environmentally as we could," Slater said when he visited the Slovenian facility. Today, you'll find ECONYL in clothing linings, dance wear, and carpets. Speedo swimsuits. Breitling watch straps. Adidas socks. Stella McCartney's Falabella handbags are lined with ECONYL jacquard, and the Falabella Go line (backpacks, overnight duffels, totes) and technical outerwear, like parkas and puffers, are made of ECONYL nylon. In late 2018, McCartney became an ECONYL supplier, too; her Italian factories collect nylon offcuts, and ECONYL reps circle round to pick up the bounty. "We love ECONYL," Bergkamp told me. "They are the only ones doing textile-to-textile on a large scale commercially."

In 2018, ECONYL accounted for 40 percent of Aquafil's production. By 2020, it should be 60 percent. And eventually Bonazzi would like it to be 100 percent. No new nylon, only recycled. Since the process "can be done an infinite number of times," he said, ECONYL is about as circular as you can get.

"You give me this carpet, I give you the yarn to make your swimwear," he said. "You give me back your swimwear, and I give you the yarn to make your jacket or, you know, the material for your 3-D printer that you are going to have at home very shortly."

WHILE THEY ARE RIGHTLY PROUD of their success, the reconstituted textiles makers want you to know that they do not "want to inadvertently bypass the reuse and repair markets," Flynn said. They only accept items "truly at the end of their life." Before garments reach that point, they all urge refurbishment and reselling—another form of circularity.

Some fashion brands make that easy. Patagonia offers consumers shopping credit to turn in old gear; it then cleans and resells the secondhand items at discount prices. The company also has a repair service called Worn Wear. Give your wounded Patagonia items to the company, and they'll be dispatched to its Reno-based rehab—the largest of its kind in North America, where forty-five full-time technicians execute fifty thousand repairs a year. "The single best thing we can do for the planet is to keep our stuff in use longer," Patagonia chief executive Rose Marcario said. "Repair reduces the need to buy more," as well as cuts "waste output and water usage."

Since 2015, Worn Wear has been on the move: menders drive around

North America in a 1991 Dodge biodiesel pickup retrofitted with a camper shell of redwood reclaimed from retired wine barrels. The menders in the truck will fix anything—from Patagonia or elsewhere—you bring to them. No charge. There is now a European version of the wagon that tools around the United Kingdom and the Continent and another in Japan.

New York designer Eileen Fisher, a longtime proponent of eco-fashion and circularity, has a take-back-and-resell program called Renew. In the last decade, the company has welcomed home more than eight hundred thousand garments—the average now is eight hundred a day—rewarding customers with a $5 shopping credit for each. Three-fourths of these garments are re-conditioned with treatments like overdyeing with pomegranate or safflower to camouflage stains and embroidering with traditional Japanese Boro and sashiko stitching, which can hide, or highlight, tears and moth holes. Renew rings up about $3 million annually.

The remaining fourth, deemed damaged beyond repair, is sent to Fisher's Tiny Factory in Irvington, New York, and her Teeny Tiny Factory, in Seattle—workshops where clothing is cut apart, rearticulated into new shapes, and sold under her label Resewn. Leftovers are fed into a felting machine and transformed into cheerful, quilt-like fabrics for throw pillows and wall hangings. And, in the summer of 2018, Fisher opened "Making Space," a hybrid concept store in Brooklyn that carries the Renew line and holds community events like sewing classes and movie screenings. "We must share," Fisher said. "We must partner in everything we do."

Some fashion leaders insist that the best way to prolong the life of clothing is to launder less, and more smartly. Levi's Chip Bergh believes you should *never* wash your jeans. "We go into autopilot," he said. "After we're finished wearing something, we just automatically toss it into the laundry." But "a good pair of denim doesn't really need to be washed . . . except for very infrequently." Washing machines beat up denim, wearing it out prematurely. Bergh suggested you do what he does: when a bit of burrito drips on your thigh, spot clean it with a toothbrush.

If you find the idea of wearing unwashed jeans appalling, and you feel the need to launder, Procter & Gamble's Bert Wouters, vice president of the global fabric enhancer franchise, counsels that you do it less and follow a simple three-step process: use a high-quality liquid or single-dose pods; run a fast cycle with cold water; finish with fabric conditioner. "When you use that regimen, you can actually extend the life of clothes by four times," he said. "And the environmental impact of that is just amazing; it's just humongous."

He's not exaggerating. If we lengthen the life of just one in five garments in Europe by 10 percent, we could cut 3 million tons of $CO_2$, save 150 million liters of water, and divert 6.4 million tons of clothing from landfill. Less frequent and shorter-cycle washing will dramatically reduce the number of microfibers flushed into our waterways. "It's actually pretty obvious when you think about it," Wouters said. "The more you put a garment to spin in a machine, the more friction there is, the more interaction there is with harsh products, the more, of course, microfibers will get released."

Through a series of initiatives, Cradle to Cradle, or C2C, has been promoting circularity practices throughout the fashion industry. In 2005, McDonough and Braungart created the Cradle to Cradle Certified Products Program and C2C Gold certification, a seal of approval determined by an independent, third-party peer review. One awardee: the C2C Gold–certified T-shirt, a recyclable organic cotton crewneck produced in 2017 by C&A and priced at €9, or about $10. "Made with wind power. The only water leaving the factory was evaporation. The people are treated with grace and dignity," McDonough explained, as he showed one of the T-shirts at the 2017 Copenhagen Fashion Summit. "It is possible."

McDonough also helped establish Fashion for Good, a global innovation platform in Amsterdam initially funded by the C&A Foundation and now counts the Ellen MacArthur Foundation and Kering among its partners.

Launched in 2017, and housed in an elegant three-story art nouveau building on the Rokin, one of Amsterdam's major thoroughfares, Fashion for Good bills itself as a "convener for change." There are two pillars to its mission, its managing director, Katrin Ley, told me: the innovation platform, with an accelerator, a scaling program, and a fashion fund; and public outreach, which includes the Fashion for Good Experience interactive museum and the Circular Apparel Community coworking and event space.

The museum, open seven days a week and free of charge, walks visitors through the various sustainability issues that dog fashion. Scary statistics are flashed on a wall, like: "Each year, 5 trillion litres of water are used for fabric dyeing alone," and "Only 5 percent of plastic waste currently is recycled." Clothes by circular brands are for sale, including that C2C Gold tee, which can be customized. The "Journey of a T-shirt" exhibit tracks the ubiquitous garment's hefty toll on the environment. The Innovation Lounge showcases circularity entrepreneurs, many of whom have passed through the accelerator or scaling programs, including Worn Again; TIPA, the Israeli company that makes compostable plastic-like biomaterial packaging; and Colorifix, a Cambridge, England–based start-up that has biologically engineered pigment that dyes fabric without the use of heavy metals, organic solvents, or acids. "If you want to create change," Ley said, "you also need to educate the wider public."

While on my tour, I ran into Worn Again's Cyndi Rhoades, sitting on a turquoise sofa in the Innovation Lounge, watching video interviews of start-up leaders on one of the museum's shiny white iPads. She was in Amsterdam for meetings with potential clients and dropped by for a chat with the scaling program's team. "No one else is doing this," she told me. "Most accelerators are B-to-B"—meaning business-to-business. "This is breaking through."

NINE

# Rage Against
# the Machine

FOR YEARS NOW, we've heard about how technology is going to radically change our relationship with clothing.

Early on, the focus was on "wearables"—the electronic devices that are embedded in garments or worn as accessories. Some have been successful, like Apple Watch and Fitbit. Some not—like, whatever happened to Google Glass? "All of our sock drawers are littered with wearable technologies where, unfortunately, the inventor didn't think it all the way through," Levi's Paul Dillinger said during my visit to the Eureka Innovation Lab.

But in recent years, fashion and science have collided in a more tangible way—in creating raw materials, as I watched at Modern Meadow and Bolt Threads; in raw material recycling, as I saw with Evrnu and its ilk; in manufacturing and distribution, via robots and automation; and in fashion design, with 3-D printing.

Three-dimensional printing has the potential to revolutionize the whole fashion game—and the change is coming sooner than you might imagine. Ray Kurzweil, a futurist who accurately predicted the exponential growth of the internet in the 1990s and the omnipresence of mobile devices in the 2000s, believes that we will be 3-D printing our own clothes by 2020. "As the variety of materials available to print in 3-D become more extensive and less expensive, both free open-source and proprietary clothing designs will be widely available online in as little as ten years," he told the *New York Times* Global Leaders' Collective conference in 2016. "It will become the norm for people to have printers in their homes."

The pitch I've heard more than once is that fashion brands—from Amazon to Chanel—won't sell you actual clothes; you'll buy a link, and you'll print the garment yourself. The possibilities are endless, as would be the shake-up to the industry as it is currently constituted. Design, production, labor, waste: it could all change beyond recognition.

How exactly would it work? Could you just print any dress? For instance, what if you had always adored Cristóbal Balenciaga's remarkable black "envelope dress"—the strapless number from 1967 that encases the body like folded paper, with four sharp corners at the shoulders? With this 3-D printing technology, could you simply zip that number out on your printer, and don it for a party? Could Balenciaga be made in Cleveland?

Andrew Bolton, the head curator of the Metropolitan Museum of Art's Costume Institute, believes 3-D printing will be "as radical as the sewing machine in terms of its democracy." It could wipe out the idea of what is old and what is new, what's in fashion and what's out. And you can forget everything you know about the business of fashion. You'll be able to make what you want, when you want it, to your size—strong-arming brands to surrender their domination of production, quality control, and distribution. Clothing design will become an exercise in file-making, in concert with other tech methods of fabrication, and bring the highest of high fashion within reach of the middle-market consumer. Anywhere.

THREE-DIMENSIONAL PRINTING, or "additive manufacturing," is not a new technology. In the early 1980s, Dr. Hideo Kodama, a Japanese lawyer at the Nagoya Municipal Industrial Research Institute, invented a rapid-prototyping technology using photopolymers that hardened in UV light to build a solid form in layers; he failed to file for the patent on time. A team of French engineers then took it up, only to be ordered by their backers, the French General Electric Company (later Alcatel Alsthom) and CILAS (the Laser Consortium) to stop, because it was not seen as "economically viable." Next came American engineer Chuck Hall, who printed a nifty eye-washing cup. He got the patent, in 1986.

Basically, 3-D printing is the construction of an item by layering molten polymer extruded in filaments or polyamide powder melted by lasers. The automobile, medical, and architecture industries all embraced 3-D printing. But fashion, with its emphasis on craft on the high end, and cheap labor on the low end, paid no attention.

That is, until Dutch couturier Iris van Herpen zeroed in.

I met "Eeee-reece," as she is known, two weeks after her thirty-fourth birthday, in Amsterdam, on the meanly cold second day of summer in 2018. We were sitting at an old wood table in her atelier—a scruffy first-floor loft in a converted warehouse at the city's former lumber port. In the nineteenth century, it was a depot for imported chocolate, cocoa, coffee, and tea. Now, it is a makers' outpost. In addition to van Herpen's headquarters, it houses studios of metalsmiths, piano tuners, and her boyfriend of nine years, Salvador Breed, a sound artist who composes haunting soundscapes for her shows.

Back in winter garb—a thick secondhand black turtleneck and jeans—and wearing little makeup and no jewelry, van Herpen reminded me of Modigliani's portraits of his wife, Jeanne Hébuterne—the wan, elongated face; the almond-shaped eyes; the rouge-blond angel hair that gently turns under at the shoulder. She speaks just as delicately.

Van Herpen has built a career on entwining fashion and technology. She honed her couture techniques as an intern for British designer Alexander McQueen in the mid-2000s, and she is forever exploring the sciences—when we met, she was reading Max Tegmark's *Life 3.0*, about artificial intelligence, and she has visited the Large Hadron Collider, the world's largest particle accelerator, at CERN, the center for particle research near Geneva, several times. "It's beautiful," she said, her eyes lighting up. "You should go. Absolutely, you should."

Her wide-ranging passions show. There are times when her designs don't even look like clothes—as least not the sort that hang in your closet or you'd wear in the street. Like the 3-D marvel that was on a mannequin next to us. Made in collaboration with Canadian architect Philip Beesley for the collection "Aeriform," it was a short, sheer shift enveloped by a cloud of paper-thin metal bubbles the size of baseballs. Picture what you would look like if you were dropped into a flute of champagne and you get the idea.

Van Herpen began experimenting with 3-D printing in 2009. With the London-based architect Daniel Widrig, she created for her "Crystallization" collection a white 3-D-printed bolero in swirling nautilus shapes. It was stone hard and encased the shoulders like a carapace. The Belgian digital manufacturer Materialise printed it from polyamide powder and took seven straight days and nights to complete it. "I thought, 'If this works, I will show it; if it doesn't work, it was a nice experiment,'" van Herpen told me. It worked.

The process, once realized, was unlike any other in fashion. "With 3-D, I could do any texture, any complexity, in any shape," she said. There was so much choice, in fact, "I felt like, 'I don't know what to do!'" But it also forced her to "decide the final look before it's made," she explained. No fittings, no trials. "The piece comes out of the machine completely finished. It's like a birth."

Excited by 3-D printing's capabilities, and possibilities, van Herpen continued to play—but always in collaboration with experts in other fields. In 2011, for her show "Escapism"—a look at our addiction to screens—she and Widrig created a superfine white 3-D–printed minidress that curled around the body like wide lacy ribbon. And for "Capriole" ("leap in the air" in French), her collection rooted in her passion for skydiving, in July 2011, she partnered with Belgian architect Isaïe Bloch and Materialise to make a bone-white "Skeleton" microdress that clung to the torso like a Day of the Dead costume.

In late 2012, van Herpen reached out to Neri Oxman, the American-Israeli architect, inventor, and designer who heads up the Mediated Matter group at the Massachusetts Institute of Technology's Media Lab, an anti-disciplinary research center that encourages commingling of disparate subjects. Oxman's field is "Material Ecology"—a term she coined for the intersection of architecture, engineering, computation, biology, and ecology.

Van Herpen explained that she wanted to make a "fully flexible" 3-D–printed dress for her next couture show in Paris in late January and asked for help. Oxman agreed. Van Herpen designed a Balenciaga-like bell-shaped miniskirt and trapeze stole, both covered with short, waxy, mono-chromatic tentacles, like those of a sea anemone. She sent sketches and photos to Oxman in Cambridge, and the MIT team set to file-making.

For two months, via Skype and email, right through the holidays, the women combined their talents to realize the outfit. Tests were printed at MIT, and the ensemble was fully executed by the Minnesota-based 3-D printing firm Stratasys. It was the first time a range of pliable materials—in this case, acrylics and polyurethane rubber—was printed in the same run. It was also the first time that color was included in the print, rather than as an aftertreatment. Entitled "Anthozoa 3-D Skirt and Cape," it is universally considered fashion's Eve of 3-D printing.

"That dress was surely a big step forward," van Herpen told me.

WHILE VAN HERPEN and Oxman were conjuring "Anthozoa" in Amsterdam and Cambridge, the Los Angeles–based jewelry, decor, and stage wear designer Michael Schmidt was designing and producing the first fully articulated 3-D–printed dress.

"I've known about 3-D printing since the 1980s," Schmidt told me. "I'd attend NASA's Technology Transfer seminars in Los Angeles, and they'd showcase science they developed for space." One of the processes was 3-D printing. "The machines were massive, and it was crude by today's standards, but you could really see this was the future," he said. "I kept my eye on the technology and waited for it to evolve."

We were sitting on old metal barstools in his studio—a treasure-cave-like space in a 1930s warehouse across the East Fourth Street Bridge from L.A.'s Arts District. The workbenches around us were cluttered with cups full of pliers and blowtorches. Tools were strategically mounted on the walls. A pair of silver plastic skeletons slouched on the sofa. Schmidt's assistants were putting finishing touches on Halloween costumes for the game show *The Price Is Right* and a pair of wax-dipped polymer rococo chandeliers for the Ace Hotel in New Orleans.

Schmidt was dressed in his usual office wear: black leather pants, a black T-shirt, heavy silver-chain bracelets of his own design, a black baseball cap on his shaggy black hair, and black motorcycle boots. His black chin-strap goatee was neatly trimmed. His twelve-year-old rescue mutt, Annie— named for his longtime friend Annie Flanders, the founding editor of *Details* magazine—napped in her bed under his desk.

A Kansas City native who barely made it through high school, Schmidt is one of the most in-demand fashion and decor designers in Los Angeles. Over the years, he has created concert tour looks for Cher, Madonna, Beyoncé, Rihanna, Janet Jackson, and Lady Gaga; stage costumes for burlesque queen Dita Von Teese; jewelry and fashion for Chrome Hearts; and metal-mesh minishifts for Jeremy Scott. Schmidt's signature style—like his look—is sexy goth-rock, with a switchblade-sharp edge.

In late 2012, one of his clients, the Ace Hotel in New York, called and asked if he would like to 3-D print a garment as the finale piece for a tech symposium they were hosting during New York Fashion Week in March.

"Yes," he said. "Now is the time to do it."

"Iris had already been doing things with 3-D printing," he told me, "and I was in love with her work. But it was more sculptural. I wanted to introduce motility, I wanted to print a fabric, which had not been done, and I wanted to create an epic moment. To do that, I needed a spectacular body and personality, and immediately thought of Dita." A striptease artist of world renown, Von Teese is credited with bringing classic burlesque back into vogue.

"I said to her, 'I'd like to create a 3-D–printed dress made to your body.' And she was super into it. So, I went back to the Ace Hotel team and said, 'I'm going to do something here that no one has ever seen before.'"

He called several Hollywood-based experts in computer-aided design—or CAD—to see if they could help write the code for the "fabric." It wouldn't be a textile, as such, but rather a plastic-like 3-D–printed nylon mesh linked together by tiny nylon rings, also printed. "I showed my designs to them and they all said it couldn't be done," he recalled. Then he reached out to Francis Bitonti, a New York–based architect and designer who uses CAD. "Francis was the first to say, 'Oh yeah, I think I can make this happen.'"

"Essentially," Schmidt explained, "the dress is a series of spirals cascading around the body." He looked to the golden ratio and the Fibonacci spiral "to quantify the ideal proportion of beauty mathematically." But the style is unquestionably his: a Maleficent-like black fishnet column with exaggerated Victorian shoulders and a plunging décolleté.

He emailed the sketch and Von Teese's measurements to Bitonti, and they collaborated via Skype. "Francis built a virtual Dita in his computer,"

Schmidt said, "and we draped the material on her in real time." Schmidt sees CAD programmers as "the new tailors" of fashion.

Bitonti's code was sent to Shapeways, a 3-D printing start-up in Queens, and the dress was made of nylon powder from the bottom up. It was printed in sections over the course of several weeks. "We started receiving these boxes of these weird shapes and flexible mesh and we had to link them together by hand," Schmidt said. "It was the most unnerving way of working I've ever done. Terrifying really, because we had one shot to get it right."

In all, the gown is made of seventeen panels, with a total of three thousand articulated joints. Everything was white; when Schmidt finished the assemblage, he dyed the dress black and decorated it with more than fifty thousand Swarovski crystals. "It needed some shimmer," he said.

To wear it, Von Teese dons a nude corset and steps into the shell, which is then laced up the back.

"The first time we had her over to try it on, I wasn't 100 percent certain that we had nailed it," Schmidt admitted. "But sure enough, it really did work. She could move in it, and walk in it, and dance in it."

Next to us, enrobing an old Stockman mannequin, was the gown. It was elegant, erotic, and looked fairly heavy. "Actually, it didn't seem heavy to me—but then I'm someone who wears sixty-five-pound costumes on stage," Von Teese told me. "It feels like nothing else I've ever worn—that's for sure. And it did wonderous things for my boobs, I must say."

When Von Teese introduced the gown at the Ace Hotel, her nearly naked, creamy-white body peeped through the diamond-shaped open weave. She also wore it for a series of pictures and a video by fashion photographer Albert Sanchez. Following the symposium, Schmidt and the dress embarked on a world tour, with stops at museums and tech expos, where he recounted its story.

He confesses the gown was a "flight of fantasy," and everyone involved donated their time and the materials. They were all game, he said, because they wanted "to see if it was even doable. Proof of concept."

What he finds amazing is how fast the technology has advanced in the intervening years. Now, "we 3-D prototype our jewelry and some sculptural work," he said. "The beauty of it is that you can create forms that are unimaginable, uncreatable in any other way."

SINCE THE VAN Herpen and Schmidt breakthroughs, 3-D printing has caught on in fashion. In 2015, London-based milliner Philip Treacy 3-D printed a series of metallic art deco–like headpieces, inspired by a hat that Greta Garbo wore in the 1931 espionage thriller *Mata Hari*. Nike, Adidas, Under Armour, and New Balance all print components of their running shoes, and some of those shoes are based on 3-D scans of customers' feet. Eyewear is 3-D printed, as are watch components, handbag closures, and shoe buckles—anything made of polymer is a candidate.

The evolution is showing no signs of slowing. "We do a lot of laser cutting combined with printing," van Herpen said. "We 3-D print molds. We 4-D print."

"4-D print?" I asked.

"You can 4-D print a flat surface, and when the surface gets warmer or wet, for example, it folds itself into the shape you want it to be."

I had seen something like this at Japanese designer Issey Miyake's Pleats Please factory in Tokyo: called Steam Stretch, the technique is a process in which a garment is made two to three times its intended size, in a polyester jacquard embedded with seams, folds, and flatish pleats. Heat is applied, and it shrinks to the correct size and shape. Then steam is applied, and it snaps into a permanent 3-D pattern: a rose blossom, or starburst, or waffling, or diamonds, or—most beautiful—a Nautilus swirl that envelops the body, like van Herpen's 3-D–printed shell.

When I described this technique to van Herpen, she brightened. "Issey is definitely one of the designers who inspires me," she said.

"For my last collection"—"Ludi Naturae," in January 2018—"I collaborated with a Dutch artist, Peter Gentenaar, who works exactly like that," in

4-D, she continued. "He takes the pulp from linen, breaks it off into micro-scopic little elements, lays it all out like a liquid, puts little lines of bamboo in it randomly; then, by applying heat, the natural material creates the shapes. It's like a metamaterial material."

"Metamaterial material?"

"Materials designed with the help of nanotechnology," she said. "You really start from the bottom up, basically."

"Atoms?"

"Yeah."

I asked about other nascent advances being discussed in these circles.

Nano-drone knitting?

Still needs tweaking, she said.

Three-dimensional printing a dress in flexible materials?

Getting there.

While what van Herpen does seems extreme, and her designs are out of reach for most consumers, she is a shining example of the good and honest side of the "cerulean-blue sweater" trickle-down system. With their high-tech experiments, she, and Miyake, and Schmidt push creativity as well as innovation in fashion. They break rules. They turn crazy dreams into real-ity. And each time they do this, technology evolves and becomes more acces-sible. Now, we can all put on a pair of 3-D printed shades.

They also encourage others to challenge the system and to modernize clothing conception, production, and consumption in a way no one dared to consider before.

And they do so not simply for the folly of it but for a just reason: the bet-terment of all.

With Anthozoa, and the Dita dress, and Steam Stretch, these out-there creators are driving the complete, and necessary, disruption of fashion.

They show us the future.

And the future is now.

EVER SINCE Richard Arkwright created the mass production model two hundred fifty years ago, there have been two guiding principles in manufacturing:

- You cannot sell a product until it is produced.
- The more you produce, the less per unit it costs—otherwise known as the "economy of scale."

Because of this, companies often produce far more than they can sell. That waste, as we've seen, is one of the chief sins of the entire fashion industry. The true test of 3-D printing has been to see if it could repudiate—and supplant—that model in a mechanized way.

Unmade is trying to do just that.

Founded in 2013, in London, by two innovation engineers and a knitwear designer, Unmade is an on-demand knitwear and print platform that facilitates the production of one-off or short-run garments and sweaters at a mass level. In short, they developed a computer program that can be grafted onto a factory's existing knitting machine and instructs said machine to knit a purple crewneck, then a red V-neck, then a black-and-white-striped scoop-neck, or whatever you—the customer—orders, one by one, without interruption. It is like what Natalie Chanin and Elizabeth Suzann are doing with their made-to-order ready-to-wear, but on an industrial level.

The implications are enormous. Since nothing is made until a customer clicks "buy" on an e-commerce site, there is no warehouse stock, no waste, no leftovers—meaning nothing needs to be marked down, or shredded, or burned. Traditional knitwear manufacturing "uses old practices and is invisible," cofounder and chief product officer Ben Alun-Jones told me in the start-up's offices in London on a late winter morning in 2017. "We are trying to open this up and make things in a more responsible, sustainable, and economically competitive way."

The company was located in a suite of windowless rooms in the bowels of Somerset House, the immense eighteenth-century former government building on the river Thames that, for the last twenty years, has served as a cultural and learning center. They were a part of Makerversity, a community of start-ups with coworking services and cut-rate rent.

Alun-Jones was twenty-nine and looked exactly how you'd imagine a tech entrepreneur to look: stalk tall and thin, in a black sweater, black jeans, black Adidas sneakers, black scruffy hair, black closely clipped beard, and round horn-rimmed glasses. The son of two doctors, he grew up in Leicestershire, which, he reminded me, had been a center for hosiery production in England until manufacturing moved offshore. He studied electrical engineering at Imperial College London.

With him was second cofounder and chief customer service officer Kirsty Emery. Thirty-one, with lake-blue eyes and falls of auburn hair, she, too, was dressed in black—tunic dress, tights, lace-up suede boots. She was raised in Northumberland, near the Scottish border, and moved to London to attend the Chelsea College of Arts, where she earned a BA in textiles, specializing in knitwear, in 2008.

Their third cofounder, Hal Watts, also thirty-one and the company's chief executive, wasn't on-site that morning. Born in Scotland, he was reared in the Paris suburb of Chantilly; his father is a commercial pilot based out of nearby Charles de Gaulle Airport. He earned his bachelor's in mechanical engineering at Imperial College London.

They all met in the late aughts while pursuing their master's degrees at the Royal College of Art—Emery in women's fashion and knitwear, Alun-Jones and Watts in innovation design engineering. Alun-Jones and Watts founded a tech design consultancy together; one of their projects was to devise a robotic piano, guitar, and drum kit for pop star will.i.am. Another was to figure out for a British brand how to customize high-performance athletic wear on an industrial level.

When they told Emery their assignment, she said: "Oh, you could knit that."

They mapped out the process and submitted it, with an estimate, to the client.

The client passed, stating the cost was too high.

"So we did it instead," Alun-Jones told me.

With the will.i.am earnings and some government grants, they developed the technology: a fashion designer would draw a sweater, which would be scanned and defined with illustration software; the digital sketch would be emailed to Unmade, or uploaded directly to its cloud; and a template would be made, "transforming the pixels into stitches," Emery said. The manufacturer would download the program into the knitting machine, produce the sweater, and ship it directly to the consumer. (They have since tweaked the system so manufacturing files are sent directly to the factory through an order management system [OMS].) There would be no need for a distribution center or a wholesaler.

Their objective was and is sustainability driven: "We want to change the planet in a positive way," Emery said. "People aren't going to stop buying clothes, so let's make what people want and will wear."

They called the company Knyttan (they changed the name to Unmade in 2015) and began by producing everything themselves in their "sample center," the large buttercup-yellow room where we were chatting. Next to us were two German-made Stoll knitting machines and one Japanese Shima Seiki machine. "These are the two big players in the knitting world," Alun-Jones told me as he showed them off; each was about twenty-five feet long and three feet wide, and they were nicknamed Hansel, Gretel, and Yoshi, respectively. "Stoll and Seiki produce 85 percent of all knitting machines, which is what manufacturers use for high apparel, footwear, T-shirts," he said.

At first, Unmade sold their wares in pop-up shops at Somerset House, in

Covent Garden, and at Selfridges department store. Then they negotiated a production deal with Johnstons of Elgin, a two-hundred-year-old mill in Hawick, Scotland's knitwear center, not far from Emery's hometown, and began to call on fashion brands. They heard the same song as Sally Fox, and Natalie Chanin, and Stacy Flynn, and Michael Schmidt.

"We were told it was impossible and we were insane," Emery recalled.

They forged ahead anyway and managed to convince the indie fashion brand Opening Ceremony and the e-commerce site Farfetch to give it a try. It was a success.

To order, you could go on the Farfetch website and search for UMd x Opening Ceremony. There would be an image of a model wearing a sweater that you could customize. (They also do scarves under their own label.) "You can change the graphic. You can change the neckline. You can add your monogram," Emery explained.

"It's like going to a tailor," Alun-Jones said. "You can put in your opinion. We are turning knitwear into a blank canvas, essentially."

When you've completed the order, the item is made and sent to you within seven to ten working days.

Unmade has since signed on several other companies, including Christopher Raeburn, Moniker, and about a half dozen "global lifestyle brands," who, Emery said, use the system in their own supply chain and prefer not to identify Unmade as the provider. Unmade charges the brands a license fee and setup costs and provides the software to the factories for free. "We are asking manufacturers to do things in a different way," Alun-Jones explained. "We want to make it as easy as possible."

We crossed the dark, narrow corridor to another big room, where a half-dozen young programmers and digital designers were working on software products and building infrastructure on desktop computers. In the back, knitwear swatches were tacked on the wall, and there was a rack of finished sweaters in Australian merino wool, cashmere, and cotton.

The company got a nice PR boost when they were included in MoMA's

*Items: Is Fashion Modern?* exhibit in 2017, where Stella McCartney's Bolt Threads dress and Modern Meadow's Zoa T-shirt premiered. For it, Unmade designed a contemporary version of the classic Breton sailor pullover, the stripes swirling paisley-like on the back. There was also an interactive touchscreen that allowed visitors to doodle a design. By the spring of 2018, Unmade had raised $4 million from investors, including London venture firms Connect Ventures, Felix Capital, and LocalGlobe. This allowed them to move to one floor up—still in the basement, still windowless, but a larger space—to Somerset House Studios, a platform and experimental workspace for tech start-ups supported by Somerset House Trust. And they've been able to hire more staff—by early 2019, they were up to thirty—to produce their software faster and more efficiently.

As I walked out of the Georgian mansion along the river Thames on that damp winter morn, I realized that I had, yet again, encountered a band of enthusiastic, idealistic, innovative Brits determined to upend the status quo in fashion. I asked Dilys Williams, of the Centre for Sustainable Fashion, why she thought so many changemakers were coming from or based in London.

"In other fashion capitals, it's about making something beautiful, very acceptable, very aesthetically pleasing," she said. "In London, it's always been about 'Fuck you.' Standing against something as well as being part of something."

H. G. WELLS, Aldous Huxley, Philip K. Dick, and Kurt Vonnegut have all noodled on whether technology is inherently evil or elicits the evil in humanity. Progressive Era economist Thorstein Veblen hailed technology, believing that it would kindle goodness in man. Trains, planes—they would carry people far and wide, and face-to-face encounters with different cultures would encourage empathy and mutual respect. But Vonnegut, who barely survived the bombing of Dresden, wasn't buying it. As he recounted in his novel *Slaughterhouse-Five*, technology can be harnessed to effect unimaginable horror. Like when we loaded planes with nuclear weapons and dropped them on Japan, killing eighty thousand in a flash. (Issey Miyake, by the way, is a Hiroshima survivor.)

The goodness of tech, like what Unmade, van Herpen, and Schmidt are doing—robots, essentially—is that, in theory, it will make redundant the hellacious jobs of Cottonopolis, Triangle Shirtwaist, Tazreen, and Rana Plaza; and that factories will be more like the tidy, quiet, AC'd Jeanologia-outfitted plant I saw in Ho Chi Minh City. But what about the Vonneguttian reality?

Throughout my reporting and research for this book, I heard every pro and every con. Robots will ax jobs that poor folk need and pummel the fledgling economies of developing nations. Robots will create better jobs, improve worker skills, and lift said economies. Robots will eliminate waste. Robots will produce more clothes than we could ever think of wearing. Robots will only take root in wealthy economies. Robots will take over everything, everywhere.

In every conversation and argument there were two consistent refrains: Robots are coming.

And they will radically change how our clothes are made and sold.

The "digital consumer is being served by an analogue supply chain," John S. Thorbeck, chairman of supply chain consultants Chainge Capital, said. Automation is "about aligning information flows."

Like 3-D printing, the notion of robots is not new. In fact, they are natural descendants of Arkwright's water frames and weavers—machines that assumed the work done for centuries by the human hand. And like 3-D printers, robots transformed the American automobile industry—first, in the early 1960s, as spot welders, and then, in the 1970s, as mobile arms capable of construction. But they certainly didn't stop there. Robots have been performing surgery for more than two decades. And they are becoming companions to humans in many realms. In the spring of 2018, I crossed an adorable semihumanoid robot nattily attired in a navy-blue blazer, yammering at passersby at the train station in Kurashiki, Japan. "Pepper," manufactured by SoftBank Robotics, is able to read emotions and is used to make people happy—thus its train station posting during rush hour.

Again, fashion was slow on the uptake. The only designer who flirted with robots was Alexander McQueen. In 1999, he closed his show "No. 13" by aiming a pair of cavorting robotic arms—the sort that spray-paint cars—at the model Shalom Harlow, who was strapped into a virginal white strapless crinoline dress, her feet lashed to the center of the set. As she spun around like a jewelry box ballerina, the metal appendages graffitied her in black and yellow. The set price was seen as performance art, not the future of clothes production.

It took the US Department of Defense to muscle the fashion industry into the bot epoch.

Back in the late aughts, a team of former researchers from Georgia Tech's Advanced Technology Development Center began to experiment with robots for sewing; they were the same brain trust who had made breakthroughs in algorithms for self-driving cars. They hypothesized that the technology that keeps cars in their lane could also guide robots to sew a straight line. They were saddened by the collapse of Georgia's apparel manufacturing industry post-NAFTA and thought that robotized, state-of-the-art factories might bring garment manufacturing back to the Peach State. Rightshoring, in a word.

In 2012, they secured a $1.26 million grant from the DoD's Defense Advanced Research Projects Agency (DARPA)—the unit that helped spawn the internet and self-driving automobiles—to "produce garments with zero direct labor." The DoD's motivation was budgetary: there are 1.3 million active-duty military members, all requiring uniforms that US law dictates must be produced domestically. American garment workers are paid by the hour; robots are not.

The team built the prototype, which attracted the attention of Palaniswamy "Raj" Rajan, an India-born, American-educated, Atlanta-based self-described "serial entrepreneur." I learned about him during my tour of Fashion for Good in Amsterdam; he was one of the rainmakers spotlighted in the

Innovation Lounge. He heads up an investment fund called CTW Venture Partners—for "change the world." Rajan chipped in a few million dollars, became chairman of the company, called SoftWear Automation, pulled in other financing, including more than a million dollars from the Walmart Foundation, and "turned it from a science project into a commercial product," he told me. He christened the automatons "Sewbots," a name he trademarked. (By 2018, DARPA had given about $2 million in total; the Walmart Foundation, $2 million; and CTW Ventures, more than $10 million.)

The Sewbots do not look human; they are boxy contraptions attached to overhead tracks, and they dart up and down the production table. To program the bots' movements, the researchers studied "how a seamstress actually operates," Rajan explained. "The first thing [sewers] do is use their eyes," and based on what they see, "they do micro and macro manipulations of the fabric with their fingers and hands and elbows and feet." The Sewbots have computer vision that analyzes the cloth, discerns where to sew, then replicates those human adjustments to stitch a seam or finish an edge.

Two years later, SoftWear installed Sewbots in a north Georgia factory and began producing bathmats and towels. "Walmart, Bed, Bath and Beyond—you buy a Made in the U.S.A. bathmat, and it'll probably be ours," Rajan said.

In 2017, SoftWear entered into a deal with Tianyuan Garments Company, a Chinese clothing manufacturer that is a major supplier for Adidas. In a brand-new $20 million plant in Little Rock, Arkansas, Sewbots run up T-shirts and parts of blue jeans. "This is not a patriotic thing," Rajan said. "It's a business decision to make locally. With today's trade wars, the Chinese chairman told me it is cheaper for him to make T-shirts in the US with robots than in China with people."

"The apparel manufacturing model as it has existed for two hundred years is fundamentally broken," Rajan explained. "You are producing goods that you don't even know if people are going to buy, and you are producing them twelve months out," on the other side of the globe.

"How do you become more efficient?" he asked. By creating "local supply chains that enable on-demand made-to-measure clothes."

As with Unmade, Sewbots can be programmed to change sewing patterns from outfit to outfit, and they are more accurate than humans—committing mistakes only 0.7 percent of the time. This all makes for less waste.

Rajan swears he will be able to get a garment from an e-commerce order to your front door in forty-eight hours if you live in close proximity to the factory, and seventy-two hours anywhere in the United States, Hawaii included. Sewbots, he predicted, will bring production back to countries that were gutted by offshoring—places like the United States, the United Kingdom, France, Japan. It's what the researchers from Georgia Tech set out to do a decade ago.

Rajan chose T-shirts and jeans as the first to be executed by Sewbots because of the amount sold each year, he said; in the US, that would be 3.5 billion T-shirts and 520 million pairs of jeans. As he pointed out, you "need a certain volume of product to be economically viable."

Volume—that word.

Volume is what gave birth to sweatshops.

Volume is what makes fast fashion so profitable.

Volume is what's stuffing our closets.

Volume is what's rotting in our landfills.

A faction in the fashion world is down on robots simply because of the potential for *explosive* volume.

By 2021, Rajan wants the Little Rock facility to be putting out twenty-four million T-shirts a year. At first. If all goes according to plan, in the next five to ten years, Rajan's Sewbots will be producing *one billion* T-shirts a year in the United States.

"People buy a lot of T-shirts," he said.

On average, ten per American, per year.

"Someone's got to make them," he said.

In 2018, SoftWear partnered with Li & Fung, the Hong Kong–based global supply chain consultancy, to initiate Sewbot production of T-shirts overseas. While Li & Fung hailed the deal as a "game-changing opportunity for manufacturers and suppliers," its group's chief executive, Spencer Fung, was more frank. "Very soon, you'll be able to pretty much automate the whole supply chain, which is pretty scary, especially for a lot of us sitting here," he told a group of fellow executives and managers shortly after the announcement.

Dilys Williams, of the Centre for Sustainable Fashion, understood what he meant. "You've got manufacturers like Li & Fung in Hong Kong talking about using more robotics to speed up the process . . . to keep producing more and more things," she said.

You've got Rajan's Sewbots zipping out bathmats every twenty seconds and T-shirts every twenty-five seconds.

You've got Nike's robots assembling running-shoe uppers twenty times faster than humans can.

You've got Adidas's "Speedfactory" in the Bavarian town of Ansbach, and a second outside of Atlanta, where shoes are cut by robots, knit by computers, and 3-D printed. "The vision is to reduce lead times from months to weeks to days or hours," Adidas tech innovation chief Gerd Manz explained.

You've got Uniqlo kitting out an entire factory with Shima Seiki 3-D knitting machines to spit out whole garments en masse.

The thought of it all caused Dilys Williams to drop her head into her hands in despair.

Then, there, in her stark office in the heart of Marylebone, as the despair began to overtake me, too, she snapped back up, looked squarely at me, and said:

"Maybe it will eat itself."


IN 2016, the International Labour Organization (ILO) predicted in its report *ASEAN in Transformation: The Future of Jobs at Risk of Automation* that up to 90 percent of workers in Southeast Asian garment factories could lose their jobs to Sewbots and other new technologies such as 3-D printing and artificial intelligence. "There are nine million sewers in the ASEAN region," the report's coauthor Jae-Hee Chang said. "People in Cambodia we spoke to basically said: 'If there are no jobs for garment workers, there's going to be another civil unrest in the country.'"

SoftWear's Rajan disputes this. "Sewbots are not economically viable in countries where wages are lower than $600 a month," he said, defensively. "Economically, it doesn't make sense to have Sewbots in Bangladesh, and this world operates on economics."

He believes in five to ten years, Sewbots will only command about 10 percent of apparel production, and by 2050, he guesses about 25 percent. And they will take over the crappy, monotonous jobs that nobody wants. "The age of cheap labor is going to end," he insisted.

The rest of apparel manufacturing will still need real live people, and he maintains they will be handling reasonable tasks in cleaner, safer environments.

Take the Little Rock factory: there are four hundred employees, and they are, by and large, highly skilled and decently paid. "People tend to the robots," he said. (On average, one human manages four bots.) There is "preprocessing, postprocessing, shipping and packing." And there is still the need for accomplished sewers to do intricate work and hand-finishing.

There will always be couture. There will always be Iris van Herpens, who only make a hundred one-of-a-kind outfits a year—special garments that convey distinct narratives. As Rajan admitted: "We'll never do a bridal dress."

But also, on the quotidian level, we have a primal urge—an instinct—to craft things by hand, and a compulsion to swathe ourselves in things made by members of our species. Anthropologists have long held forth that there are a few conditions that separate man from animals. Storytelling. Bipedalism. And the fact that we cloth ourselves. Sewing touches the human spirit.

"It's not only about the end result," van Herpen told me. "Hand-sewing focuses on something microscopic and relaxes the mind."

Technology *can* foster intimacy.

But it can also foster disengagement.

When the extraordinary tech revolution that was the internet made globalization possible and clothing production shipped overseas, "people started to consume things without actually valuing them," Dilys Williams said. "If people still had seamstresses in their families, they wouldn't be chucking things away like they do. If you don't have any idea about how something was made, you don't appreciate it."

# TEN

# To Buy or Not to Buy

BEHIND BUCKINGHAM PALACE, in the posh neighborhood of Belgravia, a private cobblestone lane is lined with
charming Victorian town houses and converted stables. One
has a wide barn door painted shell pink and is guarded by pot-
ted boxwood topiaries. Knock, and a doorman in livery will
answer.

Inside it's as if you have stepped into the dream of luxury
that marketing executives are always talking about: plush
ivory carpets, eggshell-washed walls, dusty-rose loveseats,
glass vases overflowing with peonies and spring-green vibur-
num, hushed lighting. Chic social x-rays swirl about the salons, flutes of champagne or crystal goblets of sparkling water
in hand, eyeing 24-carat-gold-plated iPads and colored-
gemstone bracelets. You might find yourself paging through a
custom clothbound edition of F. Scott Fitzgerald's *This Side of
Paradise* by the boutique bookseller Juniper Books of Colo-
rado. Or admiring a smart black leather satchel sitting on an
end table. You will be told by a polite young Englishwoman in

a prim suit that it is a Mark Cross overnight case—a re-edition of the one Grace Kelly popped open in Alfred Hitchcock's *Rear Window* to reveal to her beau Jimmy Stewart that she'd brought along her silk nightgown.

This is the Mews, a private showroom for the online luxury retailer Moda Operandi. Everything is for sale.

Founded by former *Vogue* editor Lauren Santo Domingo and Icelandic entrepreneur Áslaug Magnúsdóttir in 2010, Moda Operandi has jettisoned the traditional fashion retail model that early online players such as YOOX and Net-a-Porter still cleave to. The model—with buyers guessing what customers will like, ordering six to eight months in advance, filling warehouses with stock, and ending up saddled with leftovers that need to be marked down or discarded—has not kept pace with the evolution of e-commerce.

Instead, "Moda," as it is known to aficionados, has cooked up a retail version of how Natalie Chanin, Elizabeth Suzann, Unmade, and Rajan's Soft-Wear do business: it rings up customers' purchases based on samples, *then* orders the items, which the brands subsequently put into production. Moda embodies a business model that is both smart and fashionable and is in the vanguard of the retail revolution.

The Moda process is simple. For instance, on December 1, 2018, I went on modaoperandi.com and found a pretty pale-green one-shoulder sheath by Off-White c/o Virgil Abloh from the Resort 2019 season. The price was €831. The shipping date was April 1, 2019—four months off. If I wanted it, I could pay a 50 percent deposit: €415.50. Then the dress would be made. I'd pay the balance when it came in.

Since most of the merchandise is made to order, there are fewer leftovers threatening the environment or the bottom line. Santo Domingo said that Moda's return rate is 17 percent; by comparison, the traditional online fashion e-commerce return rate is typically 52 percent, and for preorder at department stores it is around 75 percent. This is because "we help a woman buy once and buy correctly," Santo Domingo said. "We don't encourage impulse buys." Because Moda's full-price sell-through numbers are so high, the profit margin is substantially greater than for many brick-and-mortar retailers; in 2015, Ramin Arani, of Fidelity Investments, who led Moda's fund-raising round at the time, claimed it was around a whopping 58 percent.

Santo Domingo, a lissome blond New Englander who married a Colombian beer heir, seized upon the idea for Moda while attending the fashion shows with her *Vogue* colleagues. "We would sit among the Hollywood starlets and socialites at the shows and we were able to order directly from designers," she said. "And I knew lots of women, friends of mine, would have *loved* to order directly."

A select few from the beau monde have, for decades, had that kind of access, via an old-school retail gimmick known as the "trunk show." Once the glossies had photographed the clothes, and the ad campaign had been shot, the designer would take the collection on the road. Top clients would be invited for an exclusive luncheon, tea, or cocktails, usually in-store, and see a truncated version of the catwalk show. Then they'd try on what they liked and place orders. They'd get their new clothes when the store received its seasonal delivery later in the year.

Bill Blass was the king of the trunk show. When he was in his seventies, he was still touring America's key markets, hosting at least a half-dozen himself annually. His assistants trundled out another twenty or so, in secondary markets like San Antonio, Tulsa, and Cleveland. Trunk shows benefited both brand and store—sometimes greatly: in 1993, Blass rang up more than half a million dollars during one at Saks Fifth Avenue's flagship in Manhattan—a record breaker at the time. "I'd stay out on the road for days at a time and meet people and keep my name out here," Blass said during a Nashville stop in the early 1990s. And trunk shows were like a focus group— he could gather marketing data straight from the source. "I can tell you," he reported at the time, "no matter what anyone thinks, there's a huge part of this country that still loves print dresses."

Santo Domingo thought: *Why not adapt the trunk show to the internet and bring the clothes to a broader audience, faster?*

Just as she was laying out her business plan to potential investors, social media was lifting off. Facebook and Twitter went wide in 2006, the iPhone and the app revolution came the following year, and Instagram, ultimately the key fashion platform, hit in July 2010—two months before Moda debuted online.

Until that moment, retail had been "a chain of how fashion came to the consumer: shows, retailers, the press and magazines," Robert Burke, the fashion director for Bergdorf Goodman from 1999 to 2006, told me. "Consumers were force-fed what the trends were and what they should buy. 'The Ten Must-Have Items for Fall.' There were fifty of us who would travel to New York, London, Milan, and Paris. Joan Kaner [longtime fashion director of Bergdorf's sister retailer Neiman Marcus] and I would look at slides [of runway images] and choose what we were going to buy and represent. We would have meetings with *Vogue* to recap what we were going to buy and what they were going to feature. We were able to create the trends. Women coming in with the cover of *WWD* or a picture from a magazine and saying, 'I want this'—that was a rarity."

When e-commerce and social media came up, "customers became in control of their buying," Burke said. "They decided when they wanted to shop, not be told they could do it between ten a.m. and six p.m. They had the power of what they wanted to see and deemed appropriate. And they became very well researched." Show attendees would snap with their smartphones what they saw on the runway and post the image on Twitter or Instagram. Their social media followers would scroll through the looks immediately and make their own edit, rather than waiting for retailers and glossy editors to do it for them. They'd walk into the store, open an image on their smartphone, and say, "I want this."

"This was a fundamental change," Burke said. "And it caught retailers and magazines off guard."

Since the mid-1990s, fashion executives have talked about the industry's "democratization." For the luxury end of the strata, that meant rolling out boutiques worldwide and filling them with "entry-level" logo-covered items—scarves, sunglasses, lipsticks—so anyone could afford what brand executives described as "a piece of the dream." On the mass end, it meant fast fashion, with thousands of stores stuffed full of cheap knockoffs of those pricey dreams—the cerulean-blue sweater.

But that "democratization" was not about grassroots change and taste making. The ruling class of fashion was still dictating the market to the masses; their means of production and distribution had just widened.

Social media—now *that* has truly democratized fashion. You don't have to be invited to fashion shows or private trunk shows to have an early look-see. You don't have to travel to urban centers to buy. You don't have to sub-scribe to *Vogue* or *Harper's Bazaar* to know what's on trend or learn how to wear it. Before social media, the high-fashion customer "wasn't reading blogs—please," Santo Domingo said. Today's customer "is looking at Insta-gram and following stylist Giovanna Battaglia." Admittedly, Battaglia, a former model and editor, is part of the fashion tribe. But even if that tribe still holds power, consumers—for the first time—are weighing in and making shopping decisions *before* traditional middle-market institutions, such as magazines or department stores, have made their pitch. The old retail cycle is under mortal threat. Channels of power are changing.

Americans now spend more time on digital media than working or sleep-ing, and much of that time they are looking at or buying fashion. In 2017, apparel was the number two category in US e-retail sales, after consumer electronics. Globally, fashion e-commerce hit $481.2 billion in 2018. It is ex-pected to reach about $713 billion by 2022. And not just for fast fashion. Ac-cording to McKinsey, nearly 80 percent of all luxury purchases are "digitally influenced."

Even so, folks from every demographic still love the act of shopping. Ninety-two percent of luxury buys are completed in-store. A 2017 study co-sponsored by the National Retail Federation stated that 67 percent of con-sumers under the age of twenty-one prefer hitting what they call "physical stores" rather than browsing in the ether. My own teenage daughter and her cohort much prefer bricks to clicks. And these are the kids—born on the cusp of the twenty-first century—who have never known life without Amazon. This omnichannel blend of online and bricks-and-mortar, Burke said, "is the new retail."

Thus why Santo Domingo opened the Mews in 2014. She has since added a second, called Moda Madison, on East Sixty-Fourth Street in Manhattan, and is planning a third in Hong Kong by the end of 2019. The strength of the model, known in the biz as "experiential retail," is intimacy and personalization. It's a side of retailing, Burke said, "you can't get online." Santo Domingo concurs. "The internet needs a soul, or a heart," she explained. "Otherwise, it's a headless robot."

In her cozy settings, she and her team of attractive, attentive stylists and sales associates stage live trunk shows and curate private visits featuring hard-to-find products and straight-off-the-catwalk clothes. "Last night we had a cocktail party and dinner for forty," she said as she led me through the ground-floor salon. "And it was if you were in someone's home—comfy and warm, with fires in the fireplaces."

Moda Operandi does in-home closet consultations. If you are going on vacation and want a new wardrobe for it, Moda's stylists will pull one together for you—"bathing suits, sunglasses, sundresses, sarongs, hats"— pack it up, and FedEx it to the destination, with a look book, so you know how to put each outfit together. "We can see if women from the same zip code are buying the same things. If so, we let them know and give them a chance to change their order," she said. "If two women get the same dress and wear it at the same event—that I would take to be a failure."

Such personal attention isn't solely dedicated to the *über*rich. In 2017, the Seattle-based department store chain Nordstrom introduced Nordstrom Local, an airy, ivy-covered outpost in West Hollywood, where you can go for mani-pedis, style advice, or alterations, or settle in at the tall table to work on your laptop or gab with friends over complimentary organic juice, beer, or wine. Clothes are for trying on and ordering; nothing can be taken home.

In London, Matchesfashion.com has 5 Carlos Place, a plushly carpeted, club-like affair housed in an elegant five-story Mayfair town house. The offerings on the retail floor change every two weeks or so, and like at the Mews, everything is for sale, including the contemporary art, curated and installed by the UK not-for-profit Studio Voltaire. "We're exclusive in what we offer, but inclusive, as anyone is welcome—not just the top-tier customer," one of the "style advisers" told me when I dropped in on a December Saturday. Upstairs, there are salons dedicated to themes—when I was there, it was Christmas gifting—and private suites, where you can try clothes in a spacious, flatteringly lit boudoir. On the top floor is an airy café, run by a weekly rotation of local chefs and complimentary for customers. 5 Carlos Place hosts book signings, lectures and panel discussions, master classes, and concerts and often live-streams them from an on-site studio. It's a destination experience designed to move beyond "mindless shopping," my guide said.

Burke, who since leaving Bergdorf's in the mid-2000s has run a retail consultancy in New York, sees the smaller physical footprint with a large experiential component as the future of bricks-and-mortar. "It's twelve thousand square feet versus the traditional one-hundred-thousand-square-foot department store, of which there are many. They can create a connection with the customer. They make the offerings new, fresh. Create content and buzz. I think it won't be long before Net-a-Porter and Amazon go this direction."

Santo Domingo agrees, based on what her customers tell her.

They say: "Why go to a department store that has ten floors when I can come here, with six people waiting on me, everything is my size, my taste, they know exactly what I have and don't have, everything is new? I'll never go to a department store again."

MODERN RETAILING, like modern fashion, was born in Paris in the mid-nineteenth century, when French entrepreneur Aristide Boucicaut took an existing novelty shop called Le Bon Marché and reinvented it as a mammoth, modern emporium that focused on pleasing customers, rather than soliciting them. The Bon Marché was the first to advertise extensively in newspapers. The first to fix prices. The first to accept exchanges. And refunds. And contrary to convention, customers were not pressured to buy; you could browse freely, leisurely. It was there that shopping became a jolly way for the average citizen to pass the time.

In 1869, at the height of the belle epoque, when Paris was torn up and relaid to Baron Haussmann's design, with broad, plane-tree-lined avenues and cobblestone carrefours, Boucicaut constructed a bigger, better version of the Bon Marché on the rue de Sèvres, steps off the new Boulevard Raspail. Rising four stories and topped with a milky glass atrium, it enticed customers—women, chiefly—with spectacle and richesse.

The socioeconomic theater of the Bon Marché so fascinated French author Émile Zola that, in 1883, he wrote *Au Bonheur des Dames*, or *The Ladies' Paradise*, a piquant examination of social and economic mores of the time as seen through the microcosm of the department store. "What I want to do in *The Ladies' Paradise* is write the poem of modern activity," Zola noted. "In a word, go along with the century, express the century, which is a century of action, and conquest, of effort in every direction."

The Bon Marché held the title of world's largest department store until 1914, when it was supplanted by the latest—and genuinely titanic—outpost of Chicago's bastion, Marshall Field and Company, on State Street. But Marshall Field's had been egged on to such excess by another *grande dame* of retailing, in England: Selfridges & Co.

During a vacation to London in 1906, American-born merchant Harry

Gordon Selfridge had remarked that the city lacked a dynamic department store such as Marshall Field's, where he had worked for twenty-five years. He built a six-story edifice on the central London thoroughfare of Oxford Street that took up an entire city block and boasted a half million square feet of floor. It opened in 1909.

And it didn't just sell stuff. Selfridges was a big top in the center of the city. The store displayed the first monoplane to cross the English Channel, in 1909, and twelve thousand people came to see it. In 1925, the Scottish inventor John Logie Baird introduced on the store's floor the new technology that would become television. On the roof: miniature golf and a women's gun club. *"Pull!" Pop! Pop!* So attuned was Selfridge to his customers' desires that he became known as "the chairman of shopping."

Eventually, the Great Depression, World War II, and the chairman's big-spending ways caused the store's sales and profits to plummet. Selfridge was ousted, the property was sold, and sold again; in the 1960s, it was owned by the Sears Group.

Throughout the 1960s and 1970s, American department stores were on the march. They migrated from decaying urban centers to the middle-class suburbs and exurbs, serving as anchors for shiny new shopping malls. In the 1980s and 1990s, a series of mergers and acquisitions consolidated department stores, and many of the old mainstays were lost, including Marshall Field's and the nineteenth-century workers' rights trailblazer, Wanamaker's. Both old downtown flagships now house Macy's.

And then, in turn, the dot-com revolution dealt a blow to the suburban malls. Why on Earth would you get in your car, haul off to the mall, park miles away from the store, walk more miles once inside, stand in line, hoof it all those miles back, and sit in traffic, when you could simply click on that item on a web page and have it spirited to your house?

Shoppers stopped going. The high-end retailers pulled out, and the offerings slid low. Then the stragglers began bowing out, too. In 2017 alone, an estimated 8,640 mall stores closed. And there is more to come. Credit Suisse projected in 2017 that a fourth of America's remaining malls would shutter in five years.

Even department stores' magisterial flagships on Fifth Avenue and Broadway—the ones with the glorious windows that, back in the '30s, Bill Blass spent Thursday evenings taking in—are under threat. In early 2019, the struggling retailer Lord & Taylor closed a deal to sell its century-old Italianate gem on Fifth Avenue to WeWork, the coworking company, for $850 million. The eleven-floor Midtown colossus would serve as the nine-year-old start-up's global headquarters; the first three floors would have retail, but not a Lord & Taylor.

Lord & Taylor's parent, Hudson's Bay Company, has been considering unloading some of its other historic properties—most notably the grand-daddy of them all, the century-and-a-half-old Saks Fifth Avenue in Midtown Manhattan. That pile was appraised in 2017 at $3.7 billion. Hudson's Bay shareholder Land and Buildings Investment Management suggested Amazon might be interested.

IN 2007, Amazon head Jeff Bezos reportedly told his employees, "In order to become a $200 billion company, we've got to learn how to sell clothes and food."

Exactly ten years later, he did both by acquiring the Whole Foods Market chain and launching Amazon Fashion. Yes, its clothing rubric includes unsexy staples such as socks and underwear. But there were also its new private fashion labels, like the Zara-esque Lark & Ro, and household names such as Calvin Klein, Tommy Hilfiger, and Theory. By the end of 2017, Amazon reported $175 billion in revenue, and the following year was expected to cross the $200 billion threshold. And, with projected sales of $30 billion in clothing and footwear in 2018, it was poised to overtake Walmart and Macy's and become America's number one apparel retailer.

Burke, who has consulted for Amazon, said the e-tailer plans to offer a hundred private fashion labels. Analysts predict that Amazon could be ringing up $45 to $85 billion a year in apparel sales by 2020 and possess 16 percent of the American apparel market by 2021.

To help it get there, Amazon hired Christine Beauchamp, the sunny blond former chief executive of Victoria's Secret Beauty and global brand president of Ralph Lauren's Lauren and Chaps divisions, as president of its fashion division.

Under her command, Amazon introduced in June 2018 a gadget called Echo Look: a hands-free camera and artificial intelligence personal stylist, retailing for $199. It's like having your own Mews sales assistant, but at home. Echo Look connects to Amazon's virtual assistant, Alexa; takes full-length pictures or six-second videos of you in your clothes; and builds a library of looks, which you can sort by season, style, color, or dressiness. The Style Check feature compares images of you in different outfits—like those "Who Wore It Better" magazine features—and tells you, on the screen, with the side-by-side shots, what's working, and what's not, *sans* the snark. "Fit looks better." "The shape of the outfit works better." "Colors look better on you."

Amazon also began to offer its one hundred million Amazon Prime members a try-before-you-buy service called Prime Wardrobe. (Research by Morgan Stanley shows that Prime members are twice as likely to buy fashion from the site than non-Prime members.) As with the fashion retail start-up Stitch Fix, Prime Wardrobe allows you to select a boxful of clothes, shoes, and accessories, have it delivered, and try everything in the privacy of your home. Whatever doesn't suit you can be sent back for free; what you keep, you purchase. As in other industries it has dominated in the past, Amazon's myriad of initiatives has left its peers in awe. "They're going to be massive in apparel," Levi's Chip Bergh said.

And here's why: "they want to own the brands *and* own the sourcing," Robert Burke explained to me. To that end, in 2017, Amazon secured a patent for an on-demand, automated apparel factory—meaning, like Unmade, it would produce clothes only after purchases are completed. In the patent application, Amazon declared its computerized system would provide "new ways to increase efficiency in apparel manufacturing." Fabric would be printed, cut, and sewed by a wholly automated on-site system, with cameras following the process, like a robot manager. An "image analyzer" would signal when something goes awry—the fabric gets caught in a machine, something is cut crooked—and call on a human to intervene and straighten out the situation. Fittings of finished garments would be photographed, and necessary alterations would be programmed into the production process.

"Imagine if Amazon had the body dimensions of 100 million customers," consultant and former Amazon manager James Thomson told the *Financial Times*. "Amazon doesn't replicate other people's business. They find a cheaper way to do it at a mass scale, and catch everyone going, 'What just happened?'"

IN LONDON IN October 2018, Amazon flung down its first fashion brick box. It did so as a pop-up—one of those flash shops in an unexpected spot. For five days, in a storefront near the Baker Street tube stop—a good trot from the shopping row on Oxford Street—Amazon put out autumn and winter clothes from its own labels as well as big-name brands such as Tommy Hilfiger and Calvin Klein, and, like at a fast-fashion boutique, the selection was regularly swapped out to lure customers back. Amazon's pop-up also customized Pepe jeans, staged concerts, and hosted trend talks by a British *Vogue* editor and yoga classes with a lifestyle guru.

Pop-ups were conjured by Japanese fashion avant-gardist Rei Kawakubo of Comme des Garçons in 2004. Known as "guerrilla stores," they landed in pregentrified nooks of such style hot spots as Barcelona, Singapore, and Stockholm—preferably in disused spaces, with no decor refurbishment. In Berlin, the guerrilla store was in a former bookshop; in Helsinki, in a mid-century pharmacy. The clothes were seasonless, and after a set period, the whole thing was packed up and disappeared. Economically, the guerrilla concept was very clever: rent was cheap; there was very little investment; and it drew Comme fans and the curious alike. Sales went *way* beyond expectation: the Warsaw outpost rang up three times its monthly projection in the first week.

Now, from luxury to athleisure, ephemeral retail is trending, and it's as experiential as the Mews and 5 Carlos Place. Hermès has a pop-up tour called the "Carré Club"—three- or four-day hits in cosmopolitan capitals like New York, Toronto, Los Angeles, and Milan, with everything from karaoke to a limited edition of their signature square scarves. In 2016, the French luxury brand kicked off another concept pop-up called Hermèsmatic, a mobile Laundromat that landed in second-tier markets such as Austin, Kyoto, Amsterdam, and Manchester. Customers were encouraged to bring in their old Hermès silk twill carrés—especially the ones with motifs that have aged badly—and dip-dye them in overpowering hues like beetroot, teal, and mustard for free to better effect. (How circular.) Adidas popped up in Berlin: customers could 3-D scan their bodies, then order a made-to-fit wool sweater of their own whim, ready within four hours, for €200 (about $215).

"Pop-ups have become an integral part of retail," Burke told me. "Customers today have an adverse reaction to something that is predictable—to something they feel like they've seen." Like on social media. Or on shopping websites. Pop-ups are about "newness," Burke said. "They aren't going to slow down."

Bread and circuses, à la Selfridge, are back in vogue, too. Each year, the Bon Marché spotlights a different locale—Japan, Brazil, Brooklyn, Los Angeles—jamming the store with a slew of highly curated items and larger-than-life (and slightly cliché) "animations," as the French call them. For Brooklyn, there was an old-fashioned New York–like barbershop, a pop-up tattoo shop, and "Brooklyn Amusement Park," a Cinerama-like installation by the deejays Polo & Pan of a Coney Island–like roller coaster looping its way through the borough's urban landscape. For L.A., in 2018, skateboarders tricked on a half-pipe suspended from the store's ceiling.

On Sundays, the Row DTLA, a swank new alfresco mall on the remote edges of downtown Los Angeles—exactly where American Apparel was based—hosts Smorgasburg LA, a gourmet block party that showcases its permanent vendors as well as an eclectic mix of style, vintage, and wellness merchants from around the region.

All these manifestations make for "Instagrammable" moments.

The power of Instagram in fashion cannot be overestimated. First came influencers and their accompanying hashtag #ootd, or "outfit of the day"; by late 2018, #ootd had appeared on more than two hundred million Instagram posts of influencers and their followers. Then the app begat a new pathology called the Cinderella syndrome: the resolute avoidance of being seen on social media in the same outfit twice. Cinderella syndromers order clothes online, pose in them while caching the tags, post the images, then stick them back in the box, slap on the shipping label (often a freebie), and hail UPS. Like women who used to buy a dress, wear it to a party, then return it—a sham that gave us the no tags–no return policy. Cinderella syndrome has become so extreme, brands and retailers are overwhelmed with backwash.

All of this—the pop-ups, showroom-like loci, events, social media posting—brings about what the Amazon UK fashion spokeswoman described as a "big learning experience"—tech talk for a way to harvest data.

Data is the preferred focus group for most every industry today, but it is especially true for fashion. "Every retailer needs to become data driven," said Kurt Salmon managing director Dan Murphy. It's time to stop being "instinctive and gut-driven."

Data is gleaned in all sorts of ways. Amazon, quite simply, asked visitors at the Baker Street pop-up to fill out a questionnaire. But, of course, data is collected when you make any sort of online purchase, when you click through to a site and look at merchandise. Platforms and brands use this information to sell you more stuff. Facebook/Instagram global head of luxury Morin Oluwole oversees a team in Paris that partners with brands to boost their marketing campaigns on the social media platforms, all based on data. "We can track that a consumer saw an ad on Facebook or Instagram, they clicked and went to the site, they spent XYZ amount of time, they tapped to see XYZ products and eventually purchased," she told me. "Our goal is to make sure . . . when they go to the store, they're pretty much going to buy."

Another, more Orwellian data-gathering technology is microlocating. Oluwole explained it as "a beacon [that merchants] have in their store that connects to the location services inherent in your phone" to "measure store visits." Eventually, stores will have facial recognition, so when a regular walks in, managers will be alerted—like being a savvy maître d'hôtel without having to do the homework.

One of retail's leaders "in data and customer service and marketing" is the Seattle-based department store chain Nordstrom, Burke said. In large part, this is thanks to geography. "They've hired a lot of people from Amazon," which is also headquartered in Seattle, he told me. "So their online and their data is light-years ahead of everyone else."

CONSUMERS NOW ROUTINELY and freely tell brands and retailers how they feel and what's important to them: their emotions, their envies, their neuroses. Analytics are less like taking a pulse, more like a blood draw. The fashion companies who stand above the others are those who fold such findings into a big-picture take that transcends categories like "luxury" and "athleisure" and articulates a genuine credo, like social and environmental consciousness.

And the one that embodies this best is Selfridges. The old big top has metamorphosed into the most ecologically responsible department store on Earth.

Walk in from bustling Oxford Street, and at first glance, Selfridges appears to be your typical upscale center-city bazaar. Under the spray of superbright lighting, there are gleaming cosmetics counters, charming sales assistants proffering the latest scents, shiny shelves cluttered with pricey handbags. Nothing new.

Take a closer look: those lights are LED; the signature yellow shopping bags are partly made of recycled disposable coffee cups (with an award-winning process called CupCycling); and on many of the clothes and accessories dangles a grass-green bookmark-like tag that reads "Buying Better/ Inspiring Change," a sustainability awareness campaign launched by the store in 2017. The tag explains if a product is certified organic, or has a reduced water footprint, or is British made. In 2019, the store planned to add if the items were made of "responsible leather" or were "forest friendly." To trumpet the program, there is a social media hashtag: #buyingbetter.

All this is thanks to Alannah Weston, Selfridges Group chairman and daughter of the chain's owner, the British-Canadian retail billionaire Galen Weston. Alannah joined the company in 2004, as Selfridges's creative director.

A sincere environmentalist who says she feels more comfortable in hiking boots than Louboutins, Weston initiated sustainability reforms at Selfridges's four stores, including the Oxford Street flagship, in 2011. She began with Project Ocean, a conservation program, in partnership with the Zoological Society of London, that calls attention to overfishing and destruction of marine life. Happenings included a five-week exhibition at the Oxford Street store featuring Lady Gaga's lobster hat, designed by milliner Philip Treacy, and a watery print dress from Alexander McQueen's landmark "Plato's Atlantis" collection. "When I first presented [the Project Ocean idea] to the team, they were, like, why would you do that?" Weston said. "But my dad just went straight for it. He loves anything that creates a buzz."

Three years later, Weston created a sustainability department and hired media veteran Daniella Vega to run it. A stylish brunette in her early forties, Vega spent the first ten years of her career overseeing corporate responsibility for Sky television, "using the media platform to raise issues," she explained. Weston wanted that.

Vega reached out to Dilys Williams and her team at the Centre for Sustainable Fashion to help map the store's course. All is built around the

Buying Better retail system. To put it into effect, the Centre for Sustainable Fashion and Selfridges's management organized workshops for buyers and sales assistants to "drill down into specialties, such as lingerie, menswear, beauty, and shoes, to see how and where the products were made," Vega said over breakfast in the store's Aubaine café on a warm June morning in 2017. "We wanted to build up our employees' knowledge and make sustainability another strand of how they work."

Selfridges also stages an awareness campaign each January—profiting from the public's New Year's resolution outlook—with a series of street windows calling attention to eco-ethics in fashion and design. The first, in 2016, titled "Bright New Things"—a riff on the famous 1930 English debutante class, Bright Young Things—celebrated rising British sustainability stars, such as Unmade. The following year's "Material World: What on Earth Are You Wearing?" highlighted novel materials such as yak wool and artisanal techniques like kilt pleating. "Alannah believes it's easier for a retailer to sell sustainability to customers than it is for NGOs or environmental activists," Vega explained.

In 2017, Selfridges introduced onto its digital platform the Butterfly Mark, an international interactive trust mark that offers the chain's twenty million customers each year access to a brand's supply-chain information right there on the sales floor. The Butterfly Mark was created by environmental activist Diana Verde Nieto, cofounder and CEO of Positive Luxury, a sustainable-fashion platform based in Shoreditch, London, to inform consumers world-wide on eco-conscious practices of approved brands. By early 2019, close to 150 brands, including Louis Vuitton, Temperley, Sergio Rossi, and Christian Dior, had passed the vetting process and carried the Butterfly Mark.

With the website version, you click on the butterfly that appears on the brand's or Positive Luxury's web page, and the company's environmental and social "positive actions" are listed. At Selfridges, you can see the mark on products on the store's website or smartphone app: find the item, click the butterfly, and read the supply-chain info.

Vega constantly works to make Selfridges as green as possible. She encourages designers to try new retail concepts on the store's floor, like the ECONYL wall Stella McCartney constructed to display her handbags. When Vega realized that Selfridges sold forty thousand single-serve plastic bottles of water every year in its restaurants and food halls, she banned them and had traditional water fountains installed for customers' use. In 2016, Selfridges pledged to reduce its carbon footprint 15 percent by 2020. "We are the first and only department store in the world that is Carbon Trust Triple Standard accredited," she said. "We have to make reductions in energy, waste, and water every year to maintain that standard." As she was telling me this, a waiter set my glass of orange juice on the table. It had a plastic straw in it. "That's got to go," she said, pointedly. Three months later, all plastic straws were gone.

The investment and attentiveness have paid off: in 2018, the Intercontinental Group of Department Stores named Selfridges the World's Best Department Store, a two-year honor, for the fourth time. In 2018, the company completed a £300 million, four-year renovation on the Oxford Street flagship—the greatest expenditure on a department store, ever—and posted its twelfth year of record-breaking sales, topping £1.75 billion ($2.27 billion), and an operating profit of £181 million ($234 million)—a new high. In an age of department store extinction, Selfridges is connecting brilliantly with customers. "If I look at who is doing it right, it's Selfridges," Robert Burke said.

To spread Selfridges's environmental message, Vega and her team plan to share its technology and practices with competitors and brands. Their point of view is that eco-ethics is not a marketing ploy or something that you tack onto the existing model, like adding electric window and leather upholstery options to your new car, but a holistic remake of the retailing model. "Boxing sustainability in a corner shop on the floor and not having it throughout the store—that, for me, is *not* sustainable," she says. "I dream of the day I can walk through the store and the message is in every brand I pass."

BUT WHAT IF you don't have so much money to spend on clothes? What if you have champagne tastes—you want to dress in the things you see on Moda or at Selfridges—but you are more on an H&M budget? What do you do?

I thought back to what Stella McCartney said when I initially asked her about this: "If you can't afford it the first time around, get it in the sale. Get it in the sale of the sale of the sale. Get it secondhand."

For a long time, consignment shops were filled with passé or dowdy fashion, and thrifters were down on their luck, or cash-strapped students (I loved the Salvation Army's dollar skirt rack when I was an undergrad), or counterculture. But around Year 2000, as the world was forward-obsessed, Hollywood actresses arched back into the glory years of couture for their red carpet looks, to set themselves apart from the luxury-brand-laden pack. Remember Renée Zellweger gliding into the 2001 Academy Awards in that ethereal canary-yellow chiffon Jean Dessès from 1959, purchased at Beverly Hills vintage shop Lily et Cie? Old became new again. The stigma of used was dispelled.

No one in Hollywood then, mind you, chose to wear vintage because it was green. "They were buying it because it was cool—a way for celebrities to individualize their style, just as VIP wrangling for product placement started to happen," Cameron Silver, founder and owner of the West Hollywood vintage boutique Decades, told me. "Then Hollywood and fashionable people realized the greenness of vintage—that wearing it wasn't a hand-me-down but actually a hand-me-up—and it became double-plus cool."

At the same time, the fashion production and consumption cycle was accelerating, generating a surfeit of stylish clothes—many never worn—that needed to move on.

What to do with all of that?

Consign. Online.

The queen of secondhand high fashion e-commerce is Julie Wainwright of The RealReal, an omnichannel platform based in San Francisco. Founded in 2011, The RealReal is a fashion lover's dream consignment shop: Louboutin stilettos, Louis Vuitton satchels, Gucci dresses, Chanel suits, fine jewelry, all at affordable prices. You can buy online or in one of The RealReal's boutiques; in 2018, there were two: a six-thousand-square-foot, two-floor spread on Wooster Street in Lower Manhattan, and one twice as big in West Hollywood. Consignees can reap up to a generous 85 percent of the final sale.

Wainwright had no fashion experience before The RealReal; she was a techie, and her big foray was Pets.com, which folded in 2000. But she smartly clocked millennials' dual enthusiasm for high fashion and eco-ethics. Given that consignment is, at its core, circular, The RealReal ticks both.

Wainwright's teams receive items—only from within the US as of 2018—evaluate them from "good" to "pristine," then offer them for sale on the website and in the boutiques. (She has a wily software that synchronizes what's online and in store, so a web shopper can't click and buy something that an in-person shopper is trying on in the fitting room.) "Seventy percent of our consigners have never consigned and fifty percent of our buyers have never bought consignment until they shopped with us," she said.

The RealReal's inventory at any given point is in the hundreds of thousands, and 98 percent of that stock is sold within 120 days of its posting. "We've got consumers who buy from us, then they resell the next season, and then they buy again," she said at the Copenhagen Fashion Summit in 2018. "We're getting people to circle in our own circular economy."

Wainwright bankrolled the venture with $100,000; by late 2018, she had raised $288 million. The first year, she did $10 million in turnover; in 2018, she was looking to do a billion. And it's projected to keep rising. According to a study of the US market, growth in resold items is expected to increase significantly—much faster than new items. Not surprisingly, she's considering taking The RealReal public.

Pro-green fashion groups LVMH and Kering approve of The RealReal, and Stella McCartney, as you can imagine, is 100 percent behind it. "She actually believes that if she's put in the time to make her garments—well, it should have an afterlife, should resell," Wainwright said. To encourage the practice, McCartney introduced a deal in 2018 that gives RealReal consignees of her brand a shopping credit of $100 in her stores. "Resale is a way of generating business without needing to continually create new materials and extract resources from the planet," she told me.

Some brands are still pondering how best to merge the resale model with their own business practices. As Robert Rizzolo, vice president of global merchandising for Michael Kors, acknowledged, there is a key sustainable angle to resale that should not be brushed aside. For the moment, he said, "You don't feel that when you're buying a used Chanel bag you're saving the environment. But eventually, the consumer—especially the younger consumer—is going to start to make that [connection]. It's going to increase the power of resale. . . . The future is going to be resale. We have to figure out how we can create a profit" from it. Indeed, the apparel resale market is expected to reach $41 billion by 2022.

Yet some brands regard The RealReal with suspicion, because it packs the double wallop of luxury fashion's two greatest phobias: selling online (because old-school luxury execs still consider e-commerce unsexy) and the resale of used products (because they *might* be fake). In November 2018, Chanel sued The RealReal, accusing it of passing off counterfeits as the genuine thing. In its complaint—which echoed a similar one against another vintage luxury platform, What Goes Around Comes Around—Chanel claimed the consigner, "through its business advertising and marketing practices, has attempted to deceive consumers into falsely believing that The RealReal has some kind of approval from or an association or affiliation with Chanel or that all CHANEL-branded goods sold by The RealReal are authentic."

The RealReal rejected the claim. "Chanel's lawsuit is nothing more than a thinly-veiled effort to stop consumers from reselling their authentic used goods, and to prevent customers from buying those goods at discounted prices," a spokeswoman told The Fashion Law. "They are trying to stop the circular economy."

It won't be stopped, Wainwright said, because this is what consumers believe in—the shared-values vision. "I get emails all the time from moms saying they're teaching their kids to buy on our site because it's better for the planet," she said. "We get parents . . . giving [their kids] gift cards to make their first purchase on The RealReal . . . To think differently."

AND WHAT IF YOU were to eschew the notion of ownership entirely?

When you say that all you have are the clothes on your back, it usually means you have *zero, zippo, zilch*. What if you did that with intention?

What if you bagged buying clothes and rented them instead?

Renting clothes for men is not radical. My high school boyfriend rented his tux for the prom. My husband rented his morning suit for our wedding.

But women? For anything other than for a costume party? Unthinkable. Why?

Because the fashion industry has always wanted women to *buy*.

How else could it keep the cycle going and the profits rolling in?

Then, along came the sharing economy, the idea that we don't have to possess something to consume it. We share cars, music, homes. It was only a matter of time before we shared our wardrobes.

If the internet and social media kicked off the democratization of fashion, renting solidified it. Renting clothes gives the modestly off consumer access —if briefly—to the same level of on-point luxury and style that the wealthy always have, and they can get it *right now*, for a fraction of the (oft-overinflated) retail price. The only items you'd ever need to buy are underwear, sleepwear, swimwear, and shoes. Renting, more than any other apparel business model, fulfills *Vogue* editor Anna Wintour's desire to give as many people as possible the chance to wear Fashion. And to do so daily.

Unquestionably the leader in garment sharing is the New York–based

Rent the Runway. Founded in 2009 by a pair of Harvard Business School students, Jennifer Hyman and Jennifer Fleiss, it focused for nearly its first decade in business on special-occasion clothes such as prom gowns and party garb—the nonceleb's version of red-carpet dressing.

It wasn't easy to get brands on board. Many were worried that renting could cannibalize their sales. When Rent the Runway approached New York–based Derek Lam, the label's chief executive Jan-Hendrik Schlott-mann wondered if making the clothes so available would cheapen its reputa-tion. But then he grasped that "the Rent the Runway customer is not a customer we're losing," he said. "She's not going to spend $1,500 on a dress."

No, she's the customer who goes to fast fashion and buys knockoffs. Put-ting the originals on Rent the Runway at roughly the same price could be a way to fight back.

Consider Mary Katrantzou's ongoing war with counterfeiters. I googled her name and "rent," and up popped girlmeetsdress.com, a British fashion rental site, with a few of her digital printed dresses renting for £49 (or $62). Well done, I thought. If women can rent Katrantzou's design at the same price as its rip-off—and love it so much more—maybe the rip-off market will wither and die.

In 2016—the year Rent the Runway finally turned a profit—it intro-duced a subscription service for everyday clothes, such as suits, day dresses, and outerwear from current and past seasons, as well as costume jewelry and accessories. According to Hyman, seventy-five million professional Ameri-can women shell out $3,000 or more a year buying clothes—and it's a fixed selection; you're stuck with what you have. Worse, despite the time and dough invested, there are still days you stare at your closet and feel like you have nothing to wear.

For $159 a month—or $1,908 a year—customers could retain four looks while returning what they were ready to give up and welcoming in some-thing new. (There is also a more limited $89-a-month subscription, called "RTR Update," for those with a slimmer wallet.) The cost to rent an item is

10 to 20 percent of full retail. Rent the Runway's ambition, Hyman said, is to "put Zara and H&M out of business."

It very well might. Renting fashion could change the entire shape of the apparel industry. Imagine if that $3,000-a-year clothing budget was spent on renting instead of buying. Fewer clothes would be made, and what is out there could be circulated more, tossed less. The "cerulean-blue sweater" trickle-down model would become obsolete; we'd *all* be able to wear the spiffy, sexy original rather that the poorly made, mass-marketed digestion of it.

Rent the Runway's customer base is broad, from mothers of the bride to glossy magazine editors, who—contrary to what they project—aren't that flush. One rented her wardrobe from Rent the Runway to attend Chanel's fashion show in Havana in 2016; she expensed it. Some Rent the Runway members drop by the showroom and get dressed for work or a date on the spot. With the "unlimited" subscription, they can choose a frock, scan the tag, hop in the fitting room, quick change, and be on their way. In 2018, Rent the Runway installed drop-off boxes at fifteen WeWork locations—sharing clothes in shared offices.

Customers leave comments on the site about the garments they've rented, so fellow renters know what they are getting into—another revolution in the industry. Fashion has always been supersecretive: one never revealed who made an outfit, how much it cost, if it scratched or weighed a ton or was insufferably hot or impossible to sit in. Fashion has always been about appearing effortlessly, painlessly, graciously *perfect*. Renters' reviews reveal its warts.

There are drawbacks, of course. On occasion, even after dry cleaning, garments can retain the previous wearer's natural scent—something Rent the Runway tries to mask with a floral deodorant. "The smell has really improved over time," one Unlimited subscriber told the *New Yorker*. "Though sometimes things do come and they smell like other people. You don't realize it until you heat the clothing up with your own body."

To refine how the business is run, Rent the Runway hired a chief scientist from Oracle to oversee its analytics; every phase of the rental process is determined by an algorithm. Rent the Runway also accumulates a wealth of data on rentals: from those reviews, as well as from stats like where and the number of times a customer wore an outfit. The company crafts this information into her personalized homepage, and algorithms again step in, suggesting other possibilities. Ideally, Rent the Runway would like to tighten scrolling and selection time down to two minutes. Lickety-split. Brands love this information trove, too. "It's an amazing amount of data that traditional retailers don't always share," Derek Lam's Schlottmann said. "It's great market research, frankly."

WANTING TO GIVE the rental approach a try myself, I sought out Panoply, a luxury-fashion rental company in Paris founded by two French entrepreneurs, Ingrid Brochard and Emmanuelle Brizay. I learned about Panoply from McCartney; her team reviewed several rental platforms and went with Panoply, because, her spokesman told me, it was "most like-minded."

*Oui, monsieurs et mesdames*, it is as chic as you'd imagine. The clothes are *le top*. They have panache. They have that *je ne sais quoi*. Audrey Hepburn—she would have worn any of it.

Brochard and Brizay met through mutual friends—Brochard ran MuMo, a charitable outreach program that introduced French schoolchildren in the provinces to art; Brizay had worked as a managing director for the Women's Forum for the Economy & Society—the distaff version of Davos—and managed a couple of French children's clothing brands.

They decided to go into business together. As they formed their proposal, they faced a series of questions: "How could we enable women to afford luxury?" Brizay told me. "Even for a woman who is forty, working, achieving—you cannot afford it. How do we manage to reconcile what you *want* to wear and you will actually wear?"

French women long ago mastered the art of dressing well with a tightly edited wardrobe: having a uniform of sorts, with basic, good-quality, well-cut items in neutral tones that could be changed up with a colorful scarf, or a snappy pair of shoes, or a piece of statement jewelry.

Brizay and Brochard wondered: How could they take that modest consumption model global?

They began to ponder the sharing economy.

Specifically, Brizay recalled: "If ownership is no longer what we are seeking, how does that affect fashion?"

They started in early 2016 with a pilot version among friends—"like a club"—and worked out the kinks. In November, they rolled out to the public, with a showroom on the Rue Royale, a few steps off the Place de la Concorde. And in April 2018, they added a corner at the Galeries Lafayette department store. They carry well-known labels; customers are attracted to labels they have heard of. At first, they reached out to brands; now brands reach out to them. (Chanel had recently been in touch.) They buy wholesale, like traditional retailers, but only choose a handful of looks and take only two or three sizes of each. Rental price includes delivery, pickup, and dry cleaning—primarily by Le Comptoir des Blanchisseurs, a Paris-based eco-cleaner. Delivery is by bicycle courier, Monday through Friday; otherwise, customers can collect their selection at the showroom or Galeries Lafayette. Panoply will also DHL orders throughout Europe.

I met Brizay on Panoply's second anniversary of full-time business, at the Rue Royale showroom—a former apartment in an elegant eighteenth-century building. She is quintessentially Parisienne: lean, sharp-featured, makeup-free, her dishwater-blond hair cut sassy-short. On that day, she was dressed in her own Joseph black leather pants, Nicole Farhi black cashmere pullover, and Charlotte Olympia tapestry flats, and a Panoply-rented Christopher Kane tuxedo coat. For the French, even Friday casual is the *end*.

Seated on mid-century chairs, we were surrounded by racks of clothes. Behind me were pantsuits and tuxedos by the Paris tailor Pallas (a favorite of

French first lady Brigitte Macron). In the corner, Diane von Furstenberg wrap dresses. Behind Brizay, a new arrival: dresses, suits, and coats by Stella McCartney.

One-shot clients range from teenagers looking for party dresses to eighty-year-olds who search for grandmother-of-the-bride gowns. The subscriber base is age thirty-five to fifty-five—professional women. In 2018, Panoply counted about six thousand one-shot customers and five hundred subscribers. "It's still hush-hush," Panoply's marketing and business head Sara Dalloul confided. But not too. The showroom is always humming—seven to eight customers a day.

The following week, I was attending a conference in London, at the Centre for Sustainable Fashion, and thought it would be apropos to rent my outfit. Panoply stylist Bettina Hetoubanabo, a thirty-year-old caramel-complexioned Parisienne dappled with freckles, took me in hand. I loved her look: white camisole, black jeans, black *bottines*, and mocha dreads that, at the halfway mark, segued into gold and were piled up in an ice cream swirl. "Selling is so old-fashioned," she told me, as she handed over a Stella McCartney navy wool suit with a fuchsia windowpane check. "Renting is the new way to see things."

With the boxy jacket and the cropped trousers, in such a lively plaid, the suit was not something I would have ever singled out on my own. And at the retail price of €1,720, it was far beyond what I could afford. But to rent it at €255? That was doable. I tried it, with a Phillip Lim silk camisole (€400 retail/€48 rental) that she extended. I am *not* a camisole gal. But to my surprise, the ensemble looked great. I understood what Brizay meant when she said renting clothes emboldens women—pushes them to *dare*. My personal style had amped to something decidedly more Fashion with one essay. "Our stylists take care of our clients from A to Z," Dalloul said. "If we get a client that's like, 'I have a bar mitzvah in two weeks' time, I want something to cover my arms because I don't like to show them,' our stylists will prepare a mood board and send it."

To rent, a subscriber buys a pack of credits—one credit for €69, three for €159, five for €229—then, like tickets at amusement park, doles them out as needed. (The rental price drops if you buy a bigger credit pack.) The camisole would cost me one credit; the suit three credits. The camisole, alas, was already reserved, but the suit was mine, for eight days. It was delivered by bike the next day, and off to London it went with me.

During the conference, I lapped up compliments. When I responded I'd rented it, I got more kudos. How sustainable! How circular!

In 2016, Panoply raised €1.7 million ($1.9 million), including from the investment fund Experienced Capital. And they were about to begin a second round when I met them, to double that amount. (To put it in perspective: by 2018, Rent the Runway had raised $210 million, including $20 million from Blue Pool, the capital firm that invests the wealth of Alibaba founders Jack Ma and Joe Tsai.) In 2017, Panoply bought the British rental service Chic by Choice. With it came Clean Cleaners, a sustainable dry-cleaning service, and a database of 350,000. The acquisition made Panoply the number one fashion rental company in Europe.

Brizay and Brochard have plans for further growth. Pop-ups in London. A collaboration with Air France to provide rental service to premiere clients flying from Japan to Paris, "so people can travel luggage-free," Brizay explained. A "white label," so brands can rent directly to consumers, with Panoply as the anonymous facilitator. (Rent the Runway was also looking into doing this.) Maybe even a partnership with Rent the Runway to provide a seamless service across the Atlantic. Like a global rental network.

In the spring of 2018, French prime minister Édouard Philippe unveiled the country's Circular Economy Roadmap. Among the initiatives: regulations that would prohibit fashion brands and retailers from tossing or burning unsold items. Instead, the leftovers would have to be donated to charities or recyclers (good news for Worn Again and Evrnu). Panoply sees opportunity there. Maybe they'll accept the leftovers and rent them out, too.

MANY OF THE PLAYERS in Fashionopolis are not in complete alignment. The rightshorers have an alternative aim and scale than the slow-fashion folk; high fashion will always be about seasons and, as even Stella McCartney admits, making new things. That's a different view than you'll find at companies trying to recycle and reengineer existing materials. Alabama Chanin and the Sewbots will never have the same goal, though they both could be part of the solution for the horrors exemplified by Bangladeshi sweatshops. Moda's MO is very different from Rent the Runway's.

What we have in Fashionopolis today is a complex and epic-sized mess, and it's going to take all these approaches, and many others, to tackle it and build a better, more just fashion ecosystem. Everyone spotlighted in this book is, in their way, pushing back on a model that is manifestly unsustainable. That celebrates endless consumption, ever lower prices (whether achieved by stealing someone's art or their human rights), and ever larger returns. That consciously yields leftovers. That gives no thought to the toll it takes on the environment.

The revolution is not only going to be born from the makers. We all have to step up. Buy less. Wash our clothes differently. Repair or upcycle them more. Consider the impact of the material they are made of. Consider the supply chain that produces them. Consider the tenets of the company that created and distributed them. We need to fashion a personal style that does more good for the world than ill.

And right now, renting may be about the most environmentally minded thing you can do. You can keep your look up to the minute, but the individual pieces go on to have a long life. It makes the current clothing climate *somewhat* sustainable.

Yes, renting, and all these other technologies and movements, could be bastardized—as they have been since Arkwright engaged the first water frame. Faster consumption could accelerate everything: creation, production, sales-floor drops.

Will we have even less regard for clothes if we don't invest emotion in them? If they come in and out of our lives like speed dates?

Will renting become the ultimate expression of *I really don't care, do u?*

I hope not.

I hope we will see our clothes not just as something we throw on, but as the entire ecosystem that they are.

I know that when I get dressed in the morning, I won't be casual about it anymore.

Maybe I'll put on my soft, sturdy Alabama Chanin T-shirt made of hand-picked, organic cotton, my Stony Creek indigo-dyed Levi's, held up with a Modern Meadow belt, and a smartly tailored Stella McCartney blazer made of wool from happy New Zealand sheep.

Or maybe I'll rent something saucy to step out on the town and send it back when I'm done.

And when it's all too tired to wear anymore, Stacy Flynn and Cyndi Rhoades will turn the whole lot back into virginal goodness to be woven, dyed, cut, sewed, and worn again.

The day after the sustainability conference, I returned to Paris.

And the day after that, around noon, the Panoply courier came to my flat and picked up the garment bag.

The suit was gone.

I was sad.

I liked it. A lot.

I could imagine buying it, incorporating it into my wardrobe, into my life.

Then I decided:

No, it was right for it go back.

I *could* live without it.

After all, there would always be another.

# Acknowledgments

THIS BOOK wouldn't exist without a slew of strong, smart women.

My editor, Virginia Smith Younce, at Penguin Press and my agent, Tina Bennett, at William Morris Endeavor, who believed in it, and me, and made it sing.

My research assistant, Chantel Tattoli, who helped me find the right word every time and kept me on track and carbed up.

My second assistant, Emily Wall, who tracked down obscure facts and sent missives of encouragement at *exactly* the moment they were needed.

Fact checkers Barbara Kean, Gillian Aldrich, Leslie Wiggins, and Regina Bresler, who gamely stepped in late in the game and caught my foul balls.

Photo editor Ginny Power, who has calmly pulled in pictures from all over the place for me forever.

Penguin Press assistant Caroline Sydney, whose keen eye and youthful point of view gave the manuscript the spit polish it needed.

This is a book about women, by women.

I must thank the many, many courageous folks who welcomed me into their world, told everything and then more, most notably Stella McCartney, Mary Katrantzou, Iris van Herpen, Sally Fox, Natalie Chanin, Dilys Williams, Judy Gearhart, Mar Martinez, Kalpona Akter, Sarah Bellos, Tracey Panek, Raffaella Mandriota, Tracy Hawkins, Suzanne Lee, Jamie Bainbridge, Stacy Flynn, Cyndi Rhoades, Elizabeth Pape, Emmanuelle Brizay, and especially Claire Bergkamp, who answered *so many questions*. They are all so inspiring.

Vanessa Friedman and Choire Sicha of the *New York Times*, who greenlighted and edited the first iteration of the Bangladesh reporting.

Stellene Vollandes of *Town & Country*, who initially steered me toward Lauren Santo Domingo and the Mews.

Fellow Coltrane-aphile Akram Hosen, who, on the ground in Dhaka, got me from point A to B, translated, and kept me out of trouble, and Clara Vannucci, who took the arresting photographs of our adventures. I could have never done that trip without them.

Jun Kanai and Nancy Knox Talcott for orchestrating my tour of Japan and spoiling me throughout it. I cherish their long friendship.

Michelle Finamore and her colleagues at the Museum of Fine Arts, Boston, who introduced me to the future.

Libby Callaway, and Karla Otto and her team, for setting up several key interviews.

Josh Friedman, Tom Jennings, Jonathan Logan, and everyone at the Carey Institute of Global Good in Rensselaerville, New York, who accepted me into the Logan Nonfiction Fellowship program and welcomed me to that beautiful, peaceful haven to write for three months. And instant friend-for-life Molly O'Neill, who gave me her car to get to the Y.

Hazlitt's Hotel in London and the American Library in Paris, where a lot of this book was written.

Friends who read or listened to early versions of the text, helped me get unstuck, and cheered me on when I started to flag, including Cindy Wall, Rose Apodaca, Tina Isaac, Oberon Sinclair, Teri Agins, Delores Downs, Jennifer Sullivan, my fellow Logan Fellows, Mitchell Owens, Lauren Collins, Robert Forrest, Michael Roberts, Lauren Adriana and Nick Briggs, Anthony Lane, and Luca Guadagnino.

Questlove for his epic Michelle Obama Musiaqualogy playlist, which carried me through the last three chapters.

And most importantly, Hervé and Lucie d'Halluin, who lived with the madness yet again.

They are the most courageous of all.

# Notes

INTRODUCTION

1 **Its parent company:** "Inditex's Net Sales Rise 9 Percent to €25.34 Billion in Fiscal 2017," Inditex, March 14, 2018, https://www.inditex.com/article?articleId=552792.

2 **Zara made up two-thirds:** Inditex Annual Report, 2017, 16, https://static.inditex .com/annual_report_2017/assets/pdf/memoria_en.pdf.

2 **The jacket, which came:** Gabriella Pailla, "Is Melania's Infamous Zara Jacket a Ripoff of Another Designer?" The Cut, June 22, 2018, https://www.thecut.com /2018/06/melanias-zara-jacket-r13-ripoff.html.

2 **At the time workers were:** Chase Peterson-Withorn, "The Full List of Every American Billionaire 2016," *Forbes*, March 1, 2016, https://www.forbes.com/sites /chasewithorn/2016/03/01/the-full-list-of-every-american-billionaire-2016/.

2 **Almost one kilogram:** "The Deadly Chemicals in Cotton," Environmental Justice Foundation, 2007, http://www.cottoncampaign.org/uploads/3/9/4/7/39474145 /2007_ejf_deadlychemicalsincotton.pdf.

3 **In 2013, the Center for Media:** "Brick by Brick: The State of the Shopping Center," Nielson, May 17, 2013, https://www.nielsen.com/us/en/insights/reports/2013/brick -by-brick--the-state-of-the-shopping-center.html.

3 **In 2018, that averaged sixty-eight:** Alexandra Schwartz, "Rent the Runway Wants to Lend You Your Look," *New Yorker*, October 22, 2018, https://www.newyorker .com/magazine/2018/10/22/rent-the-runway-wants-to-lend-you-your-look.

3 **the world's citizens:** Andrew Morgan, *The True Cost*, documentary (Life Is My Movie Entertainment, 2015).

3 **And if the global population:** John Kerr and John Landry, "The Pulse of the Fashion Industry," Boston Consulting Group and Global Fashion Agenda, May 2017, 8.

**3 In Tokyo:** Claire Press, "Why the Fashion Industry Is Out of Control," *Australian Financial Review*, April 23, 2016, https://www.afr.com/lifestyle/fashion/why-the-fashion-industry-is-out-of-control-20160419-goa5ic.

**4 "The expectation is to keep":** Dilys Williams, interview with the author, London, December 16, 2016.

**4 Up until the late:** Stephanie Vatz, "Why America Stopped Making Its Own Clothes," KQED News, May 24, 2013, https://www.kqed.org/lowdown/7939/madeinamerica.

**5 $2.4-trillion:** Schwartz, "Rent the Runway."

**5 In 1991, 56.2 percent:** Stephanie Clifford, "U.S. Textile Plants Return, With Floors Largely Empty of People," *New York Times*, September 12, 2013, https://www.nytimes.com/2013/09/20/business/us-textile-factories-return.html.

**5 Once-vibrant industrial:** Kate Abnett, "Does Reshoring Fashion Manufacturing Make Sense?" Business of Fashion, March 9, 2016, https://www.businessoffashion.com/articles/intelligence/can-fashion-manufacturing-come-home.

**5 The same went down:** "Global Fashion Statistics—International Apparel," Fashion United, 2018.

**5 In 2017, US apparel exports:** "Value of the U.S. Apparel Trade Worldwide from 2007 to 2017," Statista, n.d., https://www.statista.com/statistics/242290/value-of-the-us-apparel-trade-worldwide/.

**5 In 2017, Britain:** UK Fashion and Textiles Association, by email, March 4, 2019.

**5 In the summer of 2012:** Steve Denning, "Why Are the US Olympic Uniforms Being Made in China?" *Forbes*, July 23, 2012, https://www.forbes.com/sites/stevedenning/2012/07/23/why-are-the-us-olympic-uniforms-being-made-in-china.

**6 According to a 2016 poll:** Diana Verde Nieto, "What Does 'Made in America' Luxury Really Look Like?" Luxury Society, March 20, 2017, https://www.luxurysociety.com/en/articles/2017/03/what-does-made-america-luxury-really-look/.

**6 In 2018, five of:** Luisa Kroll and Kerry Dolan, "Meet the Members of the Three-Comma Club," *Forbes*, March 6, 2018, https://www.forbes.com/billionaires/#720a681d251c.

**6 Fashion employs one:** Morgan, *The True Cost.*

**6 Fewer than 2 percent:** Maxine Bédat and Michael Shank, "There Is a Major Climate Issue Hiding in Your Closet: Fast Fashion," *Fast Company*, November 11, 2016, https://www.fastcompany.com/3065532/there-is-a-major-climate-issue-hiding-in-your-closet-fast-fashion.

**6 The World Bank estimates:** Julia Jacobo, "How Sustainable Brands Are Turning Their Backs on Fast Fashion Trend," ABC News, September 13, 2016, https://abcnews.go.com/US/sustainable-brands-turning-backs-fast-fashion-trend/story?id=39590457.

**7 It releases 10 percent:** Nathalie Remy, Eveline Speelman, and Steven Swartz, "Style That's Sustainable: A New Fast-Fashion Formula," McKinsey & Company, October 2016.

**7 The fashion industry devours:** Kate Abnett, "Three Years After Rana Plaza, Has Anything Changed?" Business of Fashion, April 19, 2016, https://www.businessoffashion.com/community/voices/discussions/can-fashion-industry-become

-sustainable/-years-on-from-rana-plaza-has-anything-changed-sustainability
-safety-worker-welfare.

7 **The creation of:** Marianna Kerppola et al., "H&M's Global Supply Chain Management Sustainability: Factories and Fast Fashion," University of Michigan, February 8,
2014.

7 **World Wildlife Fund (WWF) has stated:** "The Impact of a Cotton T-Shirt," World
Wildlife Fund, January 16, 2013, https://www.worldwildlife.org/stories/the-impact
-of-a-cotton-t-shirt.

7 **Synthetic fabrics release:** Echo Huang, "The Once Pristine Waters of Antarctica
Now Contain Plastic Fibers," Quartz, June 7, 2018, https://qz.com/1299485
/antarcticas-waters-now-contain-plastic-fibers/.

7 **Of the more than 100 billion:** Alexandra Schwartz, "Rent the Runway."

7 **In the last twenty years:** Alden Wicker, "Fast Fashion Is Creating an Environmental Crisis," *Newsweek*, September 1, 2016, https://www.newsweek.com/2016/09
/09/old-clothes-fashion-waste-crisis-494824.html.

7 **The European Union disposes:** Kerr and Landry, "Pulse of the Fashion Industry," 12.

7 **In 2017, USAID:** "Overview of the Used Clothing Market in East Africa," United
States Agency for International Development, July 2017.

7 **Kenya alone:** "Global Business of Secondhand Clothes Thrive in Africa," Africa
News, April 26, 2018, http://www.africanews.com/2018/04/26/global-business-of
-secondhand-clothes-thrive-in-africa-business-africa//.

8 **In response, in 2018, the Trump:** Abdi Latif Dahir and Yomi Kazeem, "Trump's
'Trade War' Includes Punishing Africans for Refusing Second-hand American
Clothes," Quartz Africa, April 5, 2018, https://qz.com/africa/1245015/trump-trade
-war-us-suspends-rwanda-agoa-eligibility-over-secondhand-clothes-ban/.

8 **The Environmental Protection:** "Textiles: Material-Specific Data," epa.gov. n.d.

8 **In the UK, 9,513:** Press, "Why the Fashion Industry Is Out of Control."

8 **textiles are the:** Anat Keinan and Sandrine Crener, "Stella McCartney," Harvard
Business School, November 22, 2016.

8 **The National Retail Federation:** Jessica Dickler, "Black Friday Weekend: Record
$52.4 Billion Spent," CNN Money, November 27, 2011, https://money.cnn.com
/2011/11/27/pf/black_friday/index.htm.

## CHAPTER ONE: READY TO WEAR

16 **Katrantzou's fabric expert:** Raffaella Mandriota, interview with the author, Villepinte, February 13, 2018. All Mandriota quotes come from this interview.

17 **Katrantzou was born:** Mary Katrantzou, interview with the author, London, February 7, 2018. All Katrantzou quotes come from this interview, phone calls, or emails,
unless otherwise indicated.

18 **For her MA degree show:** Hanna Rose Iverson, "Started From the Bottom: Mary
Katrantzou to Hannah Griffiths," Brighton Fashion Week, n.d., http://www.brigh
tonfashionweek.com/blog/started-mary-k.

**18 in early 2018, she sold:** Samantha Conti, "Wendy Yu Invests in Mary Katrantzou, Eyes Growth in China," *Women's Wear Daily*, January 22, 2018.

**18 Weeks later, Yu:** Nancy Chilton, "Wendy Yu Endows Lead Curatorial Position at the Costume Institute," *The Metropolitan Museum of Art*, March 7, 2018.

**18 Yu earmarked:** Conti, "Wendy Yu Invests in Mary Katrantzou, Eyes Growth in China."

**20 Vogue.com:** Sarah Mower, "Mary Katrantzou," Vogue.com, September 15, 2018, https://www.vogue.com/fashion-shows/spring-2019-ready-to-wear/mary-katrantzou.

**20 The *New York Times*:** Vanessa Friedman, "The Meghan Markle Non-Effect," *New York Times*, September 18, 2018, https://www.nytimes.com/2018/09/18/fashion/london-fashion-week-spring-2019-erdem-christopher-kane.html.

**21 *Women's Wear*:** Samantha Conti, "The Collections: Mary Katrantzou," *Women's Wear Daily*, September 17, 2018.

**22 Nobody really liked Richard:** Stephen Yafa, *Cotton: The Biography of a Revolutionary Fiber* (New York: Penguin, 2005), 45.

**22 "A plain almost . . .":** Thomas Carlyle, *The Works of Thomas Carlyle* (Cambridge: Cambridge University Press, 2010), 182.

**22 The days were long:** "Working Conditions for Children," Cromfordmills.org.uk, n.d., https://www.cromfordmills.org.uk/sites/default/files/attachments/Source%202%20-%20Child%20Workers.pdf.

**22 At first, there were two:** Yafa, *Cotton*, 55.

**22 Local textile factory:** Yafa, *Cotton*, 63.

**22 By 1790, Arkwright:** Yafa, *Cotton*, 67.

**22 In 1810, a prominent:** Pietra Rivoli, *The Travels of a T-Shirt in the Global Economy: An Economist Examines the Markets, Power, and Politics of World Trade* (Hoboken, NJ: John Wiley & Sons, Inc., 2015), 96.

**22 In one of history's:** Yafa, *Cotton*, 107.

**23 Garment manufacturing:** Daniel Soyer, "Introduction: The Rise and Fall of the Garment Industry in New York City," in *A Coat of Many Colors: Immigration, Globalization, and Reform in New York City's Garment Industry*, Daniel Soyer, ed. (New York: Fordham University Press, 2005), 6–7.

**23 America's busiest port:** Hadassa Kosak, "Tailors and Troublemakers: Jewish Militancy in the New York Garment Industry, 1889-1910," in *A Coat of Many Colors: Immigration, Globalization, and Reform in New York City's Garment Industry*, Daniel Soyer, ed. (New York: Fordham University Press, 2005), 118.

**24 As the New York City:** Nancy L. Green, "From Downtown Tenements to Midtown Lofts," in *A Coat of Many Colors: Immigration, Globalization, and Reform in New York City's Garment Industry*, Daniel Soyer, ed. (New York: Fordham University Press, 2005), 34.

**24 in 1931, the New York:** "The Economic Impact of the Fashion Industry," Joint Economic Committee, United States Congress, February 6, 2015.

**24 "Everybody dressed up":** Bill Blass, "American Gals," *New Yorker*, April 14, 1997, 74.

**24** **"Black sheaths caught"**: "On and Off the Avenue: Feminine Fashions," *New Yorker*, September 20, 1941.

**25** **Joan Crawford would**: Nancy Hardin and Lois Long, "Luxury, Inc.," *New Yorker*, March 31, 1934.

**25** **Blass landed**: Blass, "American Gals."

**25** **the Garment District alone**: Soyer, "Introduction," 14–15.

**25** **Blass was one of them**: Bill Blass and Cathy Horyn, *Bare Blass* (New York: Harper-Collins, 2002), 19–21.

**25** **By the close of the 1950s**: Green, "From Downtown Tenements to Midtown Lofts," 29.

**25** **The reason was**: Soyer, "Introduction," 16.

**25** **Manhattan's apparel workers**: Soyer, "Introduction," 18.

**26** **In 1973, four hundred thousand**: Marc Karimzadeh, "Ralph Lauren Boosts N.Y. Manufacturing Initiative," *Women's Wear Daily*, October 23, 2013.

**26** **In 1965, there were**: Soyer, "Introduction," 20.

**26** **In all, 70 percent**: Stephanie Vatz, "Why America Stopped Making Its Own Clothes," KQED News, May 24, 2013.

**26** **The North American Free Trade Agreement**: "Ronald Reagan's Announcement for Presidential Candidacy," November 13, 1979, reaganlibrary.gov, https://www.reaganlibrary.gov/11-13-79.

**27** **By the late 1950s**: Alexandra Harney, *The China Price: The True Cost of Chinese Competitive Advantage* (New York: Penguin Press, 2008), 20.

**27** **Washington responded**: Rivoli, *The Travels of a T-Shirt in the Global Economy*, 160.

**27** **Even with those**: Harney, *The China Price*, 21.

**27** **In 1960, about 10 percent**: Soyer, "Introduction," 16.

**27** **"Never!" . . . Liz Claiborne**: Jerome Chazen, *My Life at Liz Claiborne: How We Broke the Rules and Built the Largest Fashion Company in the World* (Bloomington, IN: AuthorHouse, 2011), pp. 52–57.

**28** **Before long, Liz**: Chazen, *My Life at Liz Claiborne*, 67.

**28** **To achieve that**: Harney, *The China Price*, 20.

**28** **The resulting job flight**: "'Made in America' Month Bill Goes to White House," *Daily News Record*, October 15, 1986.

**28** **Unions and trade groups**: Janice H. Hammond and Maura G. Kelly, "Quick Response in the Apparel Industry," Harvard Business School, 1990.

**30** **Ortega is a lifelong**: Pankaj Ghemawat and José Luis Nueno, "Zara: Fast Fashion," Harvard Business School, December 21, 2006, 7.

**32** **NAFTA would eliminate**: "20 Things You Need to Know About NAFTA," *Baltimore Sun*, November 14, 1993.

**32** **"NAFTA means jobs"**: William F. Jasper, "From NAFTA to the NAU: NAFTA and the Security and Prosperity Partnership Are Gradual Steps toward Merging the United States, Mexico, and Canada into a North American Union," *New American*, April 16, 2007.

**32** **Texan billionaire**: "The 'Great Debate' Over NAFTA," *New York Times*, November 9, 1993.

**32 By 2006, NAFTA was:** Jasper, "From NAFTA to the NAU."

**32 In 2003, the World Bank:** Byron L. Dorgan, *Take This Job and Ship It: How Corporate Greed and Brain-Dead Politics Are Selling Out America* (New York: Thomas Dunne Books, 2006), 52.

**33 Meanwhile, from 2003:** Rivoli, *The Travels of a T-Shirt in the Global Economy*, 277.

**33 By 2000, retail spending:** Ghemawat and Nueno, "Zara: Fast Fashion," 4.

**33 In 2001, Zara had:** Ghemawat and Nueno, 8.

**33 That May, Zara's parent:** "Zara's Stunning Share Debut," BBC News, May 23, 2001, http://news.bbc.co.uk/2/hi/business/1346919.stm.

**33 Between 2001:** "#46 Zara," Forbes.com, May 2016, https://www.forbes.com/companies/zara/.

**34 But in 2015, Imran Amed:** Imran Amed and Kate Abnett, "Inditex: Agile Fashion Force," Business of Fashion, March 30, 2015, https://www.businessoffashion.com/community/voices/discussions/can-fashion-industry-become-sustainable/inditex-agile-fashion-force.

**34 "The scale was extraordinary":** Imran Amed, interview with the author, Paris, March 4, 2019.

**34 On the La Coruña campus:** Amed and Abnett, "Inditex: Agile Fashion Force."

**34 Twice a week:** Deborah Weinswig, "Retailers Should Think Like Zara: What We Learned at the August Magic Trade Show," *Forbes*, August 28, 2017, https://www.forbes.com/sites/deborahweinswig/2017/08/28/retailers-should-think-like-zara-what-we-learned-at-the-august-magic-trade-show/#3a47060e3e52.

**34 If a style doesn't:** Amed and Abnett, "Inditex: Agile Fashion Force."

**34 He announced his:** Angela Gonzaler-Rodriguez, "Amancio Ortega Retires: The Founder of Inditex Drops over 50 Executive Positions," Fashion United, December 18, 2017, https://fashionunited.com/executive/management/amancio-ortega-retires-the-founder-of-inditex-drops-over-50-executive-positions/2017121818805.

**34 *Forbes* declared he:** "Forbes Ortega Page," n.d., https://www.forbes.com/profile/amancio-ortega/.

**35 In 2017, Zara did:** Inditex Annual Report, 2017, 16.

**35 Between 2000 and 2014:** Nathalie Remy, Eveline Speelman, and Steven Swartz, "Style That's Sustainable: A New Fast-Fashion Formula," McKinsey & Company, October 2016.

**35 US consumer prices:** Chico Harlan, "The Hidden Coast of Made-in-America Retail Bargains," *Washington Post*, December 30, 2016, https://www.washingtonpost.com/news/wonk/wp/2016/12/30/the-hidden-cost-of-made-in-america-retail-bargains/?utm_term=.2134d0e18fb2.

**35 As McKinsey reported:** Remy, Speelman, and Swartz, "Style That's Sustainable."

**36 taunting us to indulge:** Marc Bain, "Fast Fashion Has Made Some of the Richest Men on Earth," Quartz, August 2, 2016, https://qz.com/747242/fast-fashion-has-made-some-of-the-richest-men-on-earth/.

**36 Fast fashion's target audience:** "Speed Is This Season's Hottest Fashion Trend," Accenture, April 5, 2017, https://newsroom.accenture.com/news/speed-is-this

-seasons-hottest-fashion-trend-according-to-research-from-kurt-salmon
-part-of-accenture-strategy.htm.

36 **"Even designer collections":** "Jean Paul Gaultier Ending RTW to Focus on Couture, Beauty," *Women's Wear Daily*, September 25, 2014.

36 **The hamster-wheel cycle:** "AP Interview: Gaultier on Madonna and Saying 'Au Revoir,'" Associated Press, April 2, 2015, https://www.apnews.com/4ae31d2efdee 43da920cc0538032268c.

36 **in 2011, *Forbes*:** "Has the Fast Fashion Cycle Created an Ex-Post Licensing System?," Fashion Law, August 13, 2015.

37 **That same year, it:** Harry Wallop, "Primark Profits Fall on High Cotton Prices," *Telegraph*, November 8, 2011, https://www.telegraph.co.uk/finance/newsbysector /retailandconsumer/8876145/Primark-profits-fall-on-high-cotton-prices .html.

## CHAPTER TWO: THE PRICE OF FURIOUS FASHION

39 **"Wow, that was quick":** Mar Martinez, interview with the author, Los Angeles, October 10, 2017. All Martinez quotes come from this interview, unless otherwise indicated.

40 **Today, Los Angeles is:** Ilse Metchek, interview with the author, Los Angeles, October 6, 2017. All Metchek quotes come from this interview, unless otherwise indicated.

40 **Martinez estimated:** Martinez, interview with the author.

41 **In 1995:** George White, "Workers Held in Near-Slavery, Officials Say," *Los Angeles Times*, August 3, 1995.

41 **According to a UCLA Labor Center:** Janna Shadduck-Hernández, Marissa Nuncio, Zacil Pech, Mar Martinez, "Dirty Threads, Dangerous Factories: Health and Safety in Los Angeles' Fashion Industry," UCLA Labor Center, 2016, https://www .labor.ucla.edu/publication/dirty-threads-dangerous-factories-health-and -safety-in-los-angeles-fashion-industry/.

41 **In 2016, the US:** Natalie Kitroeff and Victoria Kim, "Behind a $13 Shirt, a $6-an-Hour Worker," *Los Angeles Times*, August 31, 2017, https://www.latimes.com/proj ects/la-fi-forever-21-factory-workers/.

41 **Forever 21 and Ross:** Ben Bergman, "Labor Department Investigation Finds 85 Percent of LA Garment Factories Break Wage Rules," 89.3 KPCC, November 17, 2016, https://www.scpr.org/news/2016/11/17/66200/labor-department-investiga tion-finds-85-percent-of/.

42 **Most of these suppliers:** Jason McGahan, "L.A. Fashion District a Fire Trap for Garment Workers, Study Finds," *L.A. Weekly*, December 12, 2016.

43 **As *Communist Manifesto*:** *Marx Engels Collected Works*, Volume 38 (New York: International Publishers, 1846), 95, http://hiaw.org/defcon6/works/1846/letters /46_12_28.html.

44 **Mill workers:** Friedrich Engels, *The Condition of the Working Class in England* (Oxford: Oxford University Press, 1993), 75.

44 **The average life expectancy:** Stephen Yafa, *Cotton: The Biography of a Revolutionary Fiber* (New York: Penguin 2005), 103.

44 **Epidemics—cholera:** Engels, *The Condition of the Working Class in England*, 118.

44 **alcoholism surged:** Engels, 138.

44 **Half of Britain's:** Engels, 152.

44 **The endless hours:** Engels, 164–166.

44 **In the summer of 1843:** Engels, 173–175.

44 **Standing for hours:** Engels, 170–173.

45 **But the new regulations:** Engels, 182.

45 **With their philanthropic:** Engels, 283.

45 **In 1890, two young:** Nancy C. Carnevale, "Culture of Work: Italian Immigrant Women Homeworkers in the New York City Garment Industry, 1890-1914," in *A Coat of Many Colors: Immigration, Globalization, and Reform in New York City's Garment Industry*, Daniel Soyer, ed. (New York: Fordham University Press, 2005), 164.

45 **The US House:** US House of Representatives, 52nd Congress, 2nd Session, "Report of the Committee of Manufactures on the Sweating System" (Washington, DC: Government Printing Office, January 20, 1893).

45 **So activist Florence Kelley:** Eileen Boris, "Social Responsibility on a Global Level: The National Consumers League, Fair Labor, and Worker Rights at Century's End," in *A Coat of Many Colors: Immigration, Globalization, and Reform in New York City's Garment Industry*, Daniel Soyer, ed. (New York: Fordham University Press, 2005), 214–16.

46 **In 1899, the National Consumers:** Marlis Schweitzer, *When Broadway Was the Runway: Theater, Fashion, and American Culture* (Philadelphia: University of Pennsylvania Press, 2009), 75.

46 **Some retailers balked:** Boris, "Social Responsibility on a Global Level," 215.

46 **Within five years:** Boris, 215.

46 **In all, 146:** David Von Drehle, *Triangle: The Fire That Changed America* (New York: Grove Press, 2004), 167.

47 **Except the Garment District:** Blass and Horyn, *Bare Blass*, 19–22.

47 **"Manufacturers did their . . .":** Blass and Horyn, 11.

47 **Leslie Fay had long:** "Fred Pomrantz, Founder of a Women's Wear Daily," *New York Times*, February 21, 1986, https://www.nytimes.com/1986/02/21/obituaries/fred-pomrantz-founder-of-a-women-s-wear-daily.html.

48 **Fred's Wharton-educated:** Bernice Kanner, "Scandal on Seventh Avenue," *New York*, June 23, 1993.

48 **netted John Pomerantz:** Teri Agins, "Loose Threads: Dressmaker Leslie Fay Is an Old-Style Firm That's in a Modern Fix," *Wall Street Journal*, February 23, 1993.

48 **In 1986, Leslie Fay:** Kanner, "Scandal on Seventh Avenue," 40.

48 **John Pomerantz received:** Kanner, 42.

48 **the company's stock crashed:** Kanner, 47.

48 **Executives at the Wilkes-Barre:** David Silverman, "What Leslie Fay's Former CEO Learned From His Company's Bankruptcy," *Harvard Business Review*, April 20, 2010, https://hbr.org/2010/04/what-john-pomerantz-former-ceo.

**49 They learned, to their:** Bob Herbert, "In America; Leslie Fay's Logic," *New York Times*, June 19, 1994, https://www.nytimes.com/1994/06/19/opinion/in-america-leslie-fay-s-logic.html.

**50 Levi Strauss's Executive:** Lance Compa and Tashia Hinchliffe-Darricarrère, "Enforcing International Labor Rights Through Corporate Codes of Conduct," *Columbia Journal of Transnational Law*, 1995, 676, https://digitalcommons.ilr.cornell.edu/cgi/viewcontent.cgi?referer=https://www.google.com/&httpsredir=1&article=1178&context=articles.

**50 the code was "to ensure":** Bob Ortega, *In Sam We Trust: The Untold Story of Sam Walton and How Walmart Is Devouring America* (New York: Times Books, 1998), 245.

**50 Levi Strauss introduced:** Frank Swoboda, "Sears Agrees to Police Its Suppliers," *Washington Post*, March 31, 1992, https://www.washingtonpost.com/archive/business/1992/03/31/sears-agrees-to-police-its-suppliers/636ef744-6794-4780-a984-2f9c293da6c7/.

**50 The scene in the plant:** Karl Schoenberger, *Levi's Children: Coming to Terms with Human Rights in the Global Marketplace* (New York: Atlantic Monthly Press, 2000), 57.

**50 Days after Levi's:** Swoboda, "Sears Agrees to Police Its Suppliers."

**51 Eventually, the factory's:** Schoenberger, *Levi's Children*, 65.

**51 To enforce the codes:** Alexandra Harney, *The China Price: The True Cost of Chinese Competitive Advantage* (New York: Penguin Press, 2008), 197–99.

**51 In 2003, American rap stars:** "Southeast Textiles, S.A. (SETISA) Choloma, Cortes, Honduras" New York: National Labor Committee, October 2003.

**51 The industrial zone:** "Southeast Textiles, S.A. (SETISA) Choloma, Cortes, Honduras."

**51 Supervisors would "stand":** "Are We Exporting American Jobs?" Senate Democratic Policy Committee Hearing, U.S. Senate, November 14, 2003, https://www.dpc.senate.gov/hearings/hearing7/gonzales.pdf.

**52 Combs understood this news:** "Major Turn-Around at Sean P. Diddy Combs' Factory SETISA in Honduras," National Labor Committee, December 17, 2003.

**52 "Really, you work":** "Are We Exporting American Jobs?" Senate Democratic Policy Committee Hearing.

**52 In fiscal year 2018:** Ibrahim Hossain Ovi, "RMG Exports Saw 8.76 Percent Growth Last Fiscal Year," *Dhaka Tribune*, July 5, 2018, https://www.dhakatribune.com/business/2018/07/05/rmg-exports-saw-8-76-growth-last-fiscal-year.

**52 ranking Bangladesh:** Refayet Ullah Mirdha, "Bangladesh Remains the Second Biggest Apparel Exporter," *Daily Star*, August 2, 2018, https://www.thedailystar.net/business/export/bangladesh-remains-the-second-biggest-apparel-exporter-1614856.

**52 "Eighty-three percent":** Siddiqur Rahman, interview with the author, Dhaka, Bangladesh, April 21, 2018.

**53 like Judy Gearhart, the head:** Judy Gearhart, interview with the author, Washington, DC, January 25, 2017. All Gearhart quotes come from this interview, phone calls, or emails, unless otherwise indicated.

53 **But on April 11:** "Spectrum Collapse: Eight Years on and Still Little Action on Safety," cleanclothes.org, April 10, 2013, https://cleanclothes.org/news/2013/04/11/spectrum-collapse-eight-years-on-and-still-little-action-on-safety.

54 **In 2018, 10 percent:** Liana Foxvog, interview with the author, by Skype, April 13, 2018. All Foxvog quotes come from this interview, unless otherwise indicated.

54 **In December 2010:** Björn Claeson, "Deadly Secrets," International Labor Rights Forum, December 2012, 22.

54 **The scene was familiar:** Saad Hammadi and Matthew Taylor, "Workers Jump to Their Deaths as Fire Engulfs Factory Making Clothes for Gap," *Guardian*, December 14, 2010, https://www.theguardian.com/world/2010/dec/14/bangladesh-clothes-factory-workers-jump-to-death.

54 **Only then did PVH Corp.:** Scott Nova, interview with the author, by Skype, April 16, 2018.

55 **On a November evening:** Tyler McCall, "Fashioning Workers' Rights for Women," Fashionista, May 10, 2013, https://fashionista.com/2013/05/from-the-runways-of-new-york-to-the-factories-of-dhaka-fashioning-workers-rights-for-women.

55 **locked the door:** "Survivor of Bangladesh's Tazreen Factory Fire Urges U.S. Retailers to Stop Blocking Worker Safety," Democracy Now, April 25, 2013, https://www.democracynow.org/2013/4/25/survivor_of_bangladeshs_tazreen_factory_fire.

55 **Fire alarms sounded:** "Survivor of Bangladesh's Tazreen Factory Fire Urges U.S. Retailers to Stop Blocking Worker Safety."

55 **Investigators later found:** Liana Foxvog et al., "Still Waiting," Clean Clothes Campaign International Labor Rights Forum, 2013.

56 **Sohel Rana was a thug:** Jim Yardley, "The Most Hated Bangladeshi, Toppled from a Shady Empire," *New York Times*, April 30, 2013, https://www.nytimes.com/2013/05/01/world/asia/bangladesh-garment-industry-reliant-on-flimsy-oversight.html.

56 **"The crack was so huge":** Shila Begum, interview with the author, Savar, Bangladesh, April 23, 2018. All Begum quotes come from this interview.

56 **Terrified employees poured:** Yardley, "The Most Hated Bangladeshi, Toppled from a Shady Empire."

57 **Around eight a.m. the following:** Mahmudul Hassan Hridoy, interview with the author, Savar, Bangladesh, April 23, 2018. All Hridoy quotes come from this interview.

58 **At five a.m. Stockholm time:** Marianna Kerppola et al., "H&M's Global Supply Chain Management Sustainability: Factories and Fast Fashion," University of Michigan, February 8, 2014.

59 **"None of":** "H&M: Comment on Fire and Building Safety in Bangladesh," H&M, May 7, 2013.

59 **When confronted:** Foxvog et al., "Still Waiting."

60 **A slew:** Lauren McCauley, "Critics Blast US Retailers' Corporate-Dominated Factory Safety 'Sham,'" Common Dreams, July 11, 2013.

60 **Yet Americans:** Stephanie Clifford, "That 'Made in the U.S.A.' Premium," *Women's Wear Daily*, November 30, 2013.

**60 NYU's Stern Center:** Paul M. Barrett, Dorothée Baumann-Pauly, and April Gu, "Five Years After Rana Plaza: The Way Forward" (NYC Stern Center for Business and Human Rights, April 2018), 2, https://issuu.com/nyusterncenterforbusinesandhumanri/docs/nyu_bangladesh_ranaplaza_final_rele?e=31640827/64580941.

**64 He said Walmart:** Repeated attempts to contact each brand for comment or confirmation went unanswered.

**64 According to the introduction:** Attempts to contact each of Camaïeu and Roadrunner to confirm went unanswered; CJ Apparel said it did not source in Bangladesh.

**64 Tazreen's owner:** "Four Years Since the Tazreen Factory Fire: Justice Only Half Done," Clean Clothes Campaign, November 24, 2016, https://cleanclothes.org/news/2016/11/24/four-years-since-the-tazreen-factory-fire-justice-only-half-done.

**64 The trial:** Md Sanaul Islam Tipu, "No Significant Progress in Trial Six Years After Tazreen Fashions Fire," *Dhaka Tribune*, November 23, 2018, https://www.dhakatribune.com/bangladesh/2018/11/23/no-significant-progress-in-trial-six-years-after-tazreen-fashions-fire.

**64 In 2016, Sohel Rana and seventeen:** "Rana Plaza Owner, 17 Others Indicted," *Daily Star*, June 15, 2016, https://www.thedailystar.net/frontpage/rana-plaza-owner-17-others-indicted-1239742.

**64 A year later:** "Rana Plaza Owner Jailed for Three Years Over Corruption," Aljazeera.com, August 29, 2017, https://www.aljazeera.com/news/2017/08/rana-plaza-owner-jailed-years-corruption-170829161742916.html.

**66 Terrorism has:** Andrew Marszal and Chris Graham, "Twenty Hostages Killed in 'Isil' Attack on Dhaka Restaurant Popular with Foreigners," *Telegraph*, July 2, 2016, https://www.telegraph.co.uk/news/2016/07/01/gunmen-attack-restaurant-in-diplomatic-quarter-of-bangladeshi-ca/.

**66 Fashion reps immediately:** Anbarasan Ethirajan, "Islamic Attack Has Bangladesh Clothing Industry on Edge," BBC News, August 17, 2016, https://www.bbc.com/news/business-36904696.

**66 In 2016, they staged:** Maria Zimmermann, "Bangladesh: Sewing Fulltime for 61 Euros a Month," dw.com, April 23, 2017, https://www.dw.com/en/bangladesh-sewing-full-time-for-61-euros-a-month/a-38553216.

**66 All this because:** Mark Anner, interview with the author, by Skype, April 18, 2018.

## CHAPTER THREE: DIRTY LAUNDRY

**69 The average American:** Catherine Salfino, "In With the Old, In With the New," Cotton Lifestyle Monitor, January 16, 2012, https://lifestylemonitor.cottoninc.com/in-with-the-old-in-with-the-new/.

**69 buys four new pairs:** Kathleen Webber, "How Fast Fashion Is Killing Rivers Worldwide," EcoWatch, March 22, 2017, https://www.ecowatch.com/fast-fashion-river-blue-2318389169.html.

**69 "They have expression":** Emma McClendon, *Denim: Fashion's Frontier* (New York: Fashion Institute of Technology, 2016), 11.

**70 The ancient Greek historian:** Herodotus, *The History of Herodotus*, (New York: D. Appleton, 1889), 410.

**70 When the Macedonian:** James Augustin Brown Scherer, *Cotton as a World Power: A Study in the Economic Interpretation of History* (New York: Frederick A. Stokes, 1916), 6.

**70 In 63 BC, the Roman official:** Clinton G. Gilroy, *The History of Silk, Cotton, Linen, Wool, and Other Fibrous Substances* (New York: Harper and Brothers, 1845), 322.

**70 two decades later, Caesar:** *The History of Cotton* (Virginia Beach: Donning Company, 2015), 11.

**70 121.4 million bales:** James Johnson, Stephen MacDonald, Leslie Meyer, and Lyman Stone, "The World and United States Cotton Outlook," U.S. Department of Agriculture, February 23, 2018, https://www.usda.gov/oce/forum/2018/commodities/Cotton.pdf.

**70 more than one hundred:** "World Cotton Market," Cotton Australia, n.d., https://cottonaustralia.com.au/cotton-library/fact-sheets/cotton-fact-file-the-world-cotton-market.

**70 America's paper currency:** "U.S. Currency: How Money is Made—Paper and Ink," Bureau of Engraving and Printing, U.S. Department of the Treasure, n.d., https://www.moneyfactory.gov/hmimpaperandink.html.

**70 Cotton's most common use:** Stacy Howell, "How Much of the World's Clothing Is Made from Cotton?" Livestrong, July 21, 2015.

**70 Nonorganic cotton—known:** "The Deadly Chemicals in Cotton," Environmental Justice Foundation, December 31, 2007.

**70 though it is grown:** Katy Willis, "Cotton's Dirty Secret: Are You Wearing This Toxic Crop?" Timetocleanse.com, March 4, 2015, https://www.timetocleanse.com/cottons-dirty-secret/.

**70 The World Health:** Melody Meyer, "Dig Deeper, Chemical Cotton," Rodale Institute, February 4, 2014, https://www.organicconsumers.org/news/chemical-cotton.

**71 to grow one kilo:** Stephen Leahy, "World Water Day: The Cost of Cotton in Water-Challenged India," *Guardian*, March 20, 2015, https://www.theguardian.com/sustainable-business/2015/mar/20/cost-cotton-water-challenged-india-world-water-day.

**71 approximately 5,000 gallons:** Rhonda P. Hill, "1,000 Gallons of Water," Edge, August 27, 2016, https://edgexpo.com/2016/08/27/1000-gallons-of-water/.

**71 If fashion production maintains:** John Kerr and John Landry, "The Pulse of the Fashion Industry," Boston Consulting Group and Global Fashion Agenda, May 2017, 11.

**71 It was first cultivated:** Jenny Balfour-Paul, *Indigo: Egyptian Mummies to Blue Jeans* (Buffalo, NY: Firefly, 2012), 69.

**71 Like cotton, indigo:** Balfour-Paul, 60.

**72 If Strauss would cover the hefty:** Lynn Downey, *Levi Strauss: The Man Who Gave Blue Jeans to the World* (Amherst: University of Massachusetts Press, 2016), 116.

**72 Made of denim from Amoskeag:** Downey, 137.

**72 The Vault is watched over:** Tracey Panek, interview with the author, San Francisco, October 13, 2017. All Panek quotes come from this interview.

73 **standing about five feet six:** Downey, *Levi Strauss*, 141, 192.

73 **When he died in 1902:** Downey, 239.

73 **One of them, Sigmund:** Stephen Yafa, *Cotton: The Biography of a Revolutionary Fiber* (New York: Penguin, 2014 ), 110.

74 **"Jeans are sex":** Isabel Wilkinson, "The Story Behind Brooke Shields's Famous Calvin Klein Jeans," *T: The New York Times Style Magazine*, December 2, 2015, https://www.nytimes.com/2015/12/02/t-magazine/fashion/brooke-shields-calvin-klein-jeans-ad-eighties.html.

74 **The ad was so provocative:** Wilkinson, "The Story Behind Brooke Shields's Famous Calvin Klein Jeans."

74 **Klein sold:** Matt W. Cody, *Calvin Klein* (New York: Chelsea House, 2013), https://books.google.fr/books?id=xPZbAgAAQBAJ&pg=PT57&lpg=PT57&dq=calvin+klein+sales+1980&source=bl&ots=oz9gFbW98H&sig=DotejBr4jDUeIuceoohR_iMlj8E&hl=en&sa=X&redir_esc=y#v=onepage&q=calvin%20klein%20sales%201980&f=false.

74 **Jean sales rocketed:** Yafa, *Cotton*, 230.

74 **Preshrunk cotton:** Yafa, 216.

75 **The L.A.-based casualwear company:** McClendon, *Denim*, 146.

76 **By 2018, there were roughly:** Akhi Akter, "Vietnamese Textile and Apparel Industry Moving Towards US$50 Billion by 2020," *Textile Today*, February 3, 2018, https://www.textiletoday.com.bd/vietnamese-textile-apparel-industry-moving-towards-us50-billion-2020/.

76 **In 2012, jeans production:** "Production Value of Denim Jeans in Vietnam from 2012 to 2021," Statista.com, n.d., https://www.statista.com/statistics/743835/denim-jeans-production-value-vietnam/.

77 **the washhouses of Xintang:** George Ayompe, "The 'True Denim Capital of the World,' Is a Disgrace to the Industry," Aetuba.com, February 12, 2017, https://www.linkedin.com/pulse/true-denim-capital-world-disgrace-industry-we-should-act-ayompe/.

78 **as cotton expert Sally Fox:** Sally Fox, interview with the author, Brooks, California, October 14, 2017. All Fox quotes come from this interview, phone calls or emails, unless otherwise indicated.

80 **when the brand was reporting:** Caroline Fairchild, "Does Levi Strauss Still Fit America?" *Fortune*, October 6, 2014, http://fortune.com/2014/09/18/levi-strauss-chip-bergh/.

80 **For much of the twentieth century:** Karl Schoenberger, *Levi's Children: Coming to Terms with Human Rights in the Marketplace* (New York: Atlantic Monthly Press, 2000), 25–31.

81 **"It is more than paternalism":** Robert Howard, "Values Make the Company: An Interview with Robert Haas," *Harvard Business Review*, October 1990, https://hbr.org/1990/09/values-make-the-company-an-interview-with-robert-haas.

81 **Levi's started selling at discounters:** Teri Agins and Joann S. Lublin, "Levi Strauss Sees Philip Marineau As a Right Fit to Be Brand Builder," *Wall Street Journal*, September 8, 1999, https://www.wsj.com/articles/SB93674980745078128.

**81 Levi's announced it would close eleven:** Nina Munk, "How Levi's Trashed a Great American Brand," *Fortune*, April 12, 1999, http://archive.fortune.com/maga zines/fortune/fortune_archive/1999/04/12/258131/index.htm.

**81 citing high labor costs:** "Spotlight: Belgian Strike Protests Levi Strauss Plant Closings," *Los Angeles Times*, October 2, 1998.

**81 Annabelle Nichols, a straight-backed:** Annabelle Nichols, interview with the author, Nashville, August 23, 2016. All Nichols quotes come from this interview.

**81 The Knoxville facility:** Suzanne Coile, "Levi's Strauss Plant," *Our Life at Work*, December 6, 2012.

**81 It opened in 1953:** "Back in the Day: Halting Those Denim Blues," *Knoxville News Sentinel*, October 5, 2014, http://archive.knoxnews.com/business/back-in-the-day -halting-those-denim-blues-ep-648080710-354171531.html.

**81 each about the size:** Coile, "Levi's Strauss Plant."

**82 On Monday, November 3, 1997:** "Levi to Close 11 Plants, Cut a Third of Jobs," Associated Press, November 4, 1997.

**82 Levi's swore it wasn't:** "Levi to Close 11 Plants, Cut a Third of Jobs."

**82 "dislocated workers":** Coile, "Levi's Strauss Plant."

**82 That year alone, Levi's:** Schoenberger, *Levi's Children*, 162.

**83 Levi's sales continued:** Amy Doan, "Levi Strauss' Frayed Fortunes," *Forbes*, September 29, 2000.

**83 The company announced more plant closings:** Schoenberger, *Levi's Children*, 54.

**83 To execute the closures:** Agins and Lublin, "Levi Strauss Sees Philip Marineau as a Right Fit to Be Brand Builder."

**83 That meant shutting down:** Jenny Strasburg, "Appalachian Travails," *San Francisco Chronicle*, June 16, 2002, https://www.sfgate.com/business/article/APPALA CHIAN-TRAVAILS-A-tiny-Georgia-town-faces-2808292.php.

**83 Most workers earned $8:** Fred Dickey, "Levi Strauss and the Price We Pay," *Los Angeles Times Magazine*, December 1, 2002.

**83 Some took home as little as:** Strasburg, "Appalachian Travails."

**84 The state opened an employment:** Strasburg, "Appalachian Travails."

**84 In all, he sacked:** "Levi Strauss & Co/On the Record: Phil Marineau," *San Francisco Chronicle*, March 5, 2006, https://www.sfgate.com/business/ontherecord/article /LEVI-STRAUSS-CO-On-the-Record-Phil-Marineau-2521804.php.

**84 In the midst of firing:** "Levi's Phil Marineau Could Net $4 Million," *Women's Wear Daily*, July 15, 2005.

**84 At the close of those:** Michael Liedtke, "Levi's CEO to Step Down by End of Year," *Orange County Register*, July 6, 2006, https://www.ocregister.com/2006/07 /06/levi-strauss-ceo-to-step-down-after-seven-rocky-years/.

**86 Monsanto introduced:** Yafa, *Cotton*, 278.

**86 in 2018, 94 percent:** "Adoption of Genetically Engineered Crops in the U.S.," United States Department of Agriculture, July 16, 2018, https://www.ers.usda.gov /data-products/adoption-of-genetically-engineered-crops-in-the-us.aspx.

**86 Roundup is the world's:** Charles M. Benbrook, "Trends in the Glyphosate Herbicide Use in the United States and Globally," U.S. National Library of Medicine: National Institutes of Health, February, 2, 2016, https://www.ncbi.nlm.nih.gov /pmc/articles/PMC5044953/.

86 **In 1994, Patagonia founder:** Yafa, *Cotton*, 291.

86 **in 2015, the International:** Michael Specter, "Roundup and Risk Assessment," *New Yorker*, April 10, 2015, https://www.newyorker.com/news/daily-comment /roundup-and-risk-assessment.

86 **Burglars in South:** "Dog Poisoning with the Intention to Break into Houses," South Africa Today, July 10, 2014, https://southafricatoday.net/south-africa-news/dog -poisoning-with-the-intention-to-break-into-houses/.

87 **"the pressure of capital markets":** David Weil, interview with the author, Boston, September 28, 2017. All Weil quotes come from this interview.


CHAPTER FOUR: FIELD TO FORM

91 **"They used cotton":** Natalie Chanin, interview with the author, Florence, Alabama, August 25, 2016. All Chanin quotes come from this interview, phone calls, or emails, unless otherwise indicated.

92 **"NAFTA destroyed":** Terry Wylie, by email, August 24, 2018.

92 **His signature look is:** Mike Albo, "Dressed to Impress, with a Southern Drawl," *New York Times*, February 18, 2009, https://www.nytimes.com/2009/02/19/fashion /19CRITIC.html.

94 **She sees her:** Kristi York Wooten, "You Can Make It There," Bitter Southerner, n.d., https://bittersoutherner.com/alabama-chanin.

94 **"My mother":** Debbie Elliott, "Reviving a Southern Industry, From Cotton Field to Clothing Rack," *NPR Morning Edition*, October 10, 2014, https://www.npr.org /2014/10/10/354934991/reviving-a-southern-industry-from-cotton-field-to -clothing-rack.

96 **"It was artistic":** Julie Gilhart, interview with the author, by phone, July 29, 2018.

97 **"the nurturing benefits":** Wooten, "You Can Make It There."

101 **Born in 1964:** Billy Reid, interview with the author, Florence, Alabama, August 27, 2016.

101 **His Spring-Summer 2002:** Rachel Dodes, "Designer Fashions a Comeback Without the Usual Pattern," *Wall Street Journal*, February 9, 2006.

102 **"Being in Florence":** K.P. McNeill, interview with the author, Florence, Alabama, August 28, 2016. All McNeill quotes come from this interview, phone calls, or emails, unless otherwise indicated.

103 **With his vertically run:** Dodes, "Designer Fashions a Comeback Without the Usual Pattern."

103 **Three years after:** Jean E. Palmieri, "Billy Reid Reflects on 20 Years in Fashion," *Women's Wear Daily*, December 28, 2017.

103 **"We broke down":** Elliott, "Reviving a Southern Industry."

103 **K.P. McNeill was driving:** Elliott, "Reviving a Southern Industry."

104 **They reached out:** Rinne Allen, "Billy Reid and Alabama Chanin's Homegrown Cotton," *T: The New York Times Style Magazine*, May 9, 2014, https://tmagazine

.blogs.nytimes.com/2014/09/05/billy-reid-and-alabama-chanin-sustainable
-cotton-project/.

104 **"So many people"**: Elliott, "Reviving a Southern Industry."

105 **The mill owner**: Allen, "Billy Reid and Alabama Chanin's Homegrown Cotton."

105 **Locklear—known as**: Steven Kurutz, "The Sock Queen of Alabama," *New York Times*, March 29, 2016, https://www.nytimes.com/2016/03/31/fashion/sock
-business-alabama.html.

105 **In 2008, at the age**: Gina Locklear, interview with the author, by phone, September 20, 2018. All Locklear quotes come from this interview, unless otherwise indicated.

106 **about seven hundred**: Elliott, "Reviving a Southern Industry."

107 **In 2017, the Nashville Fashion**: Lizzy Alfs, "Report: Nashville Fashion Industry Contributes Billions to Economy," *Tennessean*, January 25, 2017, https://eu.tennessean
.com/story/money/2017/01/25/report-nashville-fashion-industry-contributes
-billions-economy/97011064/.

107 **More than half**: Lauren Sherman, "Nashville: America's Next Fashion Capital?" Business of Fashion, April 4, 2017, https://www.businessoffashion.com/articles
/market-gps/nashville-americas-next-fashion-capital.

107 **had been established**: Alfs, "Report: Nashville Fashion Industry Contributes Billions to Economy."

108 **A self-taught sewer**: Elizabeth Pape, interview with the author, Nashville, August 25, 2016. All Pape quotes come from this interview, unless otherwise indicated.

109 **At Omega Apparel**: Nichols, interview with the author, Nashville.

109 **In early 2020**: David Perry, interview with the author, by phone, September 27, 2018.

## CHAPTER FIVE: RIGHTSHORING

113 **a mill town**: Friedrich Engels, *The Condition of the Working Class in England* (Oxford: Oxford University Press, 1993).

114 **"we built a modern"**: Tracy Hawkins, interview with the author, Dukinfield, UK, November 23, 2016. All Hawkins quotes come from this interview, phone calls, or emails, unless otherwise indicated.

114 **the first large-scale**: Paul Byrne, "Cotton to Be Spun in UK Mill for the First Time in 30 Years," *Mirror*, December 2, 2015, https://www.mirror.co.uk/news/uk-news
/cotton-spun-uk-mill-first-6944260.

114 **With a private investment**: "King Cotton Comes Home," Innovation Textiles, December 3, 2015, https://www.innovationintextiles.com/king-cotton-comes-home
-manchester-company-invests-58m-to-bring-cotton-spinning-back-to-britain/.

116 **McCormack and Shaughnessy**: Shelina Begum, "Cotton Spinning Returns to Greater Manchester Thanks to £5.8m Investment," *Manchester Evening News*, December 5, 2015, https://www.manchestereveningnews.co.uk/business/business-news
/cotton-spinning-returns-greater-manchester-10532083.

116 **English Fine Cottons produced**: Ashley Armstrong, "Riding High," *Sunday Telegraph*, December 10, 2017.

117 **In the United States**: Patrick Van den Bossche et al., "The Truth About Reshoring," A.T. Kearney, 2014.

**117 the second-fastest-growing:** Kate Abnett, "Does Reshoring Fashion Manufacturing Make Sense?" Business of Fashion, March 9, 2016, https://www.businessoffash ion.com/articles/intelligence/can-fashion-manufacturing-come-home.

**117 an impressive vault:** Rebecca Mead, "The Garmento King," *New Yorker*, September 23, 2013, https://www.newyorker.com/magazine/2013/09/23/the-garmento-king.

**117 In Great Britain:** Abnett, "Does Reshoring Fashion Manufacturing Make Sense?"

**117 and another 20,000:** Karen Kay, "Luxury Brands Feel Demand for Return of UK's Cotton and Knitwear Mills," *Guardian*, October 30, 2016, https://www.theguardian .com/fashion/2016/oct/30/fashion-luxury-brands-return-of-uk-cotton-mills.

**117 Paul Donovan:** Natasha Turak, "We May Have Hit 'Peak Trade' Even Without Trump's Tariffs, Economist Says," CNBC, March 7, 2018, https://www.cnbc.com /2018/03/07/we-may-have-hit-peak-trade-even-without-trumps-tariffs-ubs.html.

**118 Rightshoring has so:** Katie Weisman, "Made in the USA: Dead or Alive?" Business of Fashion, November 13, 2017, https://www.businessoffashion.com/articles/intelli gence/made-in-the-usa-dead-or-alive.

**118 In South Carolina:** Stephanie Clifford, "U.S. Textile Plants Return, With Floors Largely Empty of People," *New York Times*, September 12, 2013.

**119 In 2015, the Zhejiang-based:** "Keer Group to Invest $218 Million to Create 501 Jobs in Lancaster Country," South Carolina Department of Commerce, December 16, 2013, https://www.sccommerce.com/news/keer-group-invest-218-million-create -501-jobs-lancaster-county.

**119 there are "incentives":** Jenni Avins, "Chinese Textile Manufacturers Found a Cheap New Place for Outsourcing: The US," Quartz, August 4, 2015, https://qz.com /470358/chinese-textile-manufacturers-found-a-cheap-new-place-for -outsourcing-the-us/.

**119 As Lancaster County:** Hiroko Tabuchi, "Chinese Textile Mills Are Now Hiring in Places Where Cotton Was King," *New York Times*, August 2, 2015, https://www .nytimes.com/2015/08/03/business/chinese-textile-mills-are-now-hiring-in-places -where-cotton-was-king.html.

**120 Designer Nanette Lepore:** Stephanie Clifford, "That 'Made in the U.S.A.' Premium," *Women's Wear Daily*, November 30, 2013.

**120 what she calls:** Maria Cornejo, interview with the author, New York, June 28, 2017. All Cornejo quotes come from this interview, phone calls, or emails, unless otherwise indicated.

**121 On her first day:** Maura Egan, "Maria Cornejo, the Independent," *New York Times*, February 8, 2014, https://www.nytimes.com/2014/02/08/t-magazine/maria-cornejo -style.html.

**123 The borough already:** Vanessa Friedman, "Brooklyn's Wearable Revolution," *New York Times*, April 30, 2016.

**123 During his State:** Arthur Friedman, "An Apparel Campus Grow in Brooklyn," *Women's Wear Daily*, February 14, 2017.

**123 New York City had 1,568:** Winnie Hu, "New York Tries to Revive Garment Industry, Outside the Garment District," *New York Times*, February 7, 2017, https://www .nytimes.com/2017/02/07/nyregion/new-york-garment-industry-brooklyn.html.

**123 designer Yeohlee Teng:** Jean E. Palmieri, "Fashion Insiders Spar Over N.Y.'s Garment District Location," *Women's Wear Daily*, April 24, 2017.

**123 Joe Ferrara:** Valeriya Safronova, "A Debate Over the Home of New York's Fashion Industry," *New York Times*, April 25, 2017, https://www.nytimes.com/2017/04/25/fashion/de-blasio-garment-district-sunset-park.html.

**123 Like costume:** Safronova, "A Debate Over the Home of New York's Fashion Industry."

**124 Rosen dissents:** Andrew Rosen, interview with the author, New York, June 22, 2017. All Rosen quotes come from this interview or phone calls, unless otherwise indicated.

**126 she says she wants:** Dhani Mau, "A Look Inside Reformation's Bright, Shiny, Sustainable Los Angeles Factory," Fashionista, April 25, 2017, https://fashionista.com/2017/04/reformation-factory.

**126 Fans include:** John Koblin, "Reformation, an Eco Label the Cool Girls Pick," *New York Times*, December 17, 2014.

**126 and Meghan Markle:** "Meghan Markle Just Wore a Dress with a Thigh-High Slit and Looked Incredible," Cosmopolitan.com, October 22, 2018, https://www.cosmopolitan.com/entertainment/a24056950/meghan-markle-thigh-high-slit-dress/.

**126 "I design for":** Kristen Bateman, "How Reformation Because the Ultimate Cool Girl Brand for Sustainable Clothes," *Allure*, February 15, 2017, https://www.allure.com/story/reformation-yael-aflalo-sustainable-fashion-brand.

**126 As early as 2004:** Jenny Strasburg, "Made in the U.S.A.," *San Francisco Chronicle*, July 4, 2004, https://www.sfgate.com/business/article/MADE-IN-THE-U-S-2709678.php.

**127 In 1999, at twenty-one:** Koblin, "Reformation, an Eco Label the Cool Girls Pick."

**128 "I want altruism":** Koblin, "Reformation, an Eco Label the Cool Girls Pick."

**128 She took over:** Danielle Directo-Meston, "Inside Reformation's Sustainable Sewing Factory and HQ in Boyle Heights," Racked, March 19, 2015, https://la.racked.com/2015/3/19/8227687/reformation-downtown-los-angeles-studio.

**128 She bought a Tesla:** Kathleen Chaykowski, "This Model Turned CEO Is Betting 'Bricks and Clicks' Can Create a Green Fast-Fashion Empire," *Forbes*, October 24, 2017.

**128 (Few have.):** Yael Aflalo, interview with the author, by phone, October 12, 2018. All Aflalo quotes come from this interview, unless otherwise indicated.

**128 "We make killer":** Emily Holt, "Meet the Woman Behind Cool Ethical Label Reformation," *Vogue*, November 4, 2015, https://www.vogue.com/article/reformation-eco-fashion-ethical-label.

**128 she hired Zara's trend:** Koblin, "Reformation, an Eco Label the Cool Girls Pick."

**128 Board member Ken Fox:** Chaykowski, "This Model Turned CEO."

**128 "The prevailing sustainable":** Holt, "Meet the Woman."

**129 Aflalo adopted:** Chaykowski, "This Model Turned CEO."

**130 She told us:** Kathleen Talbot, interview with the author, Los Angeles, October 13, 2017.

**131 Aflaflo believed:** Chaykowski, "This Model Turned CEO."

CHAPTER SIX: MY BLUE HEAVEN

135 **"We were out":** Sarah Bellos, interview with the author, Goodlettsville, TN, August 23, 2016. All Bellos quotes come from this interview, phone calls, or emails, unless otherwise indicated.

137 **"Crazy," she remarked:** Amy Feldman, "Stony Creek Colors Is Convincing Tobacco Farmers to Grow Indigo, Building a Business on Natural Dyes," *Forbes*, August 27, 2017, https://www.forbes.com/sites/forbestreptalks/2017/08/27/stony-creek-colors-is-convincing-tobacco-farmers-to-grow-indigo-building-a-business-on-natural-dyes/.

138 **The EPA classifies aniline:** Jasmin Malik Chua, "Axing Aniline in Denim Dyes? Not So Fast," *Sourcing Journal*, December 3, 2018, https://sourcingjournal.com/denim/denim-mills/axing-aniline-129031/.

138 **Recent reports:** Melody M. Bomgardner, "These New Textile Dyeing Methods Could Make Fashion More Sustainable," *Chemical & Engineering News*, July 15, 2018, https://cen.acs.org/business/consumer-products/new-textile-dyeing-methods-make/96/i29.

139 **Tobacco is finicky:** Emily Siner, "Why Tobacco Farmers in Robertson County Are Switching to Indigo," Nashville Public Radio, August 22, 2016, https://www.nashvillepublicradio.org/post/why-tobacco-farmers-robertson-county-are-switching-indigo#stream/0.

139 **Susceptible to disease:** Feldman, "Stony Creek Colors Is Convincing Tobacco Farmers."

140 **Bellos had no trouble:** Feldman, "Stony Creek Colors Is Convincing Tobacco Farmers."

141 **"Our job is to":** Young Lee, "David Hieatt, Founder of Hiut Denim," Heddels, November 9, 2013, https://www.heddels.com/2013/11/david-hieatt-founder-hiut-denim-exclusive-interview/.

142 **In 1965, Kotaro:** Paul Travi, "Big John—The History of the First Japanese Made Jeans," Heddels, March 26, 2013, https://www.heddels.com/2013/03/big-john-the-history-of-the-first-japanese-made-jeans/.

143 **Soon, 70 percent of all:** Emma McClendon, *Denim: Fashion's Frontier* (New York: Fashion Institute of Technology, 2016), 31.

143 **In the 1980s, Japanese "pickers":** McClendon, *Denim*, 30.

143 **In Osaka:** "The History of the Osaka 5," Heddels, March 17, 2014.

143 **One company:** McClendon, *Denim*, 31.

144 **On a misty:** Tatsushi Tabuchi, interview with the author, Kojima, Japan, April 5, 2018.

147 **"Jeanologia was born":** Enrique Silla, interview with the author, Valencia, Spain, December 19, 2016. All Silla quotes come from this interview.

148 **about half of what:** Malcom Moor, "The End of China's Cheap Denim Dream," *Telegraph*, February 26, 2011, https://www.telegraph.co.uk/news/worldnews/asia/china/8349425/The-end-of-Chinas-cheap-denim-dream.html.

150 **"Keynesian economics":** Paul Dillinger, interview with the author, San Francisco, October 13, 2017. All Dillinger quotes come from this interview, or phone calls, unless otherwise indicated.

**150 A vegan who:** "Chip Bergh: 'There Are No Real Failures—Only Opportunities to Learn," Thrive Global, January 26, 2017, https://medium.com/thrive-global/chip -bergh-there-are-no-real-failures-only-opportunities-to-learn-e8972dd96b73.

**150 runs marathons:** Adam Bryant, "Chip Bergh on Setting a High Bar and Holding People Accountable," *New York Times*, June 9, 2017, https://www.nytimes.com /2017/06/09/business/chip-bergh-on-setting-a-high-bar-and-holding-people -accountable.html.

**150 Bergh promised:** Russell Hotten, "How Jeans Giant Levi Strauss Got Its Mojo Back," BBC News, September 25, 2017, https://www.bbc.com/news/business-40945709.

**150 He replaced ten:** Tessa Love, "Levi's CEO Chip Bergh Leads Company Rebound in Part by Winning Battle for Executive Talent," *San Francisco Business Times*, November 11, 2016, https://www.bizjournals.com/sanfrancisco/news/2016/11/10/most -admired-chip-bergh-levi-strauss-rebound.html.

**150 Curleigh figured:** James Curleigh, speech, VAMFF Business Seminar, Melbourne, Australia, March 17, 2015.

**151 At the time, Levi's:** Tim Higgins, "Distressed Denim: Levi's Tries to Adapt to the Yoga Pants Era," *Bloomberg Businessweek*, July 23, 2015.

**151 whenever the design:** Caroline Fairchild, "Does Levi Strauss Still Fit America?" *Fortune*, October 6, 2014, http://fortune.com/2014/09/18/levi-strauss-chip-bergh/.

**151 "We probably . . ."** Higgins, "Distressed Denim."

**151 In 2012, he decided:** Bart Sights, interview with the author, San Francisco, October 13, 2017. All Sights quotes come from this interview, unless otherwise indicated.

**153 A Dubliner:** Largetail, "Interview: Paul O'Neill of Levi's Vintage Clothing," Cool-hunting, February 17, 2015, https://coolhunting.com/style/interview-paul-oneill -levis-vintage-clothing/.

**153 Four months later:** Mark Lane, "Levi Strauss Replaces People with Lasers," Apparel Insider, March 1, 2018, https://apparelinsider.com/levi-strauss-replaces-people -lasers/.

**154 In 2003, Cone Mills:** David Shuck, "Who Killed the Cone Mills White Oak Plant?" Heddels, February 1, 2018, https://www.heddels.com/2018/02/killed-cone -mills-white-oak-plant/.

**154 In October 2016:** Shuck, "Who Killed the Cone Mills White Oak Plant?"

**154 laying off the last:** Alex Williams, "No Room for America Left in Those Jeans," *New York Times*, November 10, 2017, https://www.nytimes.com/2017/11/10/style/good bye-american-selvage-jeans.html.

**155 A couple of weeks:** Sabrina Simms, "Sunday Focus: What Are the Next Steps for Vidalia Denim?" *Natchez Democrat*, July 22, 2018, https://www.natchezdemocrat .com/2018/07/22/sunday-focus-what-are-the-next-steps-for-vidalia-denim/.

**155 It would source sustainable:** Dan Feibus, interview with the author, by phone, October 11, 2018. All Feibus quotes come from this interview.

**156 Wrangler was one:** "Vidalia Denim to Supply Sustainably-Made Denim Fabrics from State-of-Art Facility in Louisiana," Market Insider, July 24, 2018, https://mar kets.businessinsider.com/news/stocks/vidalia-denim-to-supply-sustainably-made -denim-fabrics-from-state-of-art-facility-in-louisiana-1027394624.

CHAPTER SEVEN: WE CAN WORK IT OUT

**160 A lifelong vegetarian:** Anat Keinan and Sandrine Crener, "Stella McCartney," Harvard Business School, November 22, 2016, 19.

**160 McCartney believes:** Stella McCartney, interview with the author, London, March 16, 2017. All McCartney quotes come from this interview, unless otherwise indicated.

**160 Millennial and:** Elizabeth Doupnik, "Ath-Leisure-Clad, Sustainably Aware Consumers Catapult Wool Market," *Women's Wear Daily*, March 27, 2017, https://wwd .com/fashion-news/textiles/spotlight-woolmark-10851043/.

**161 a Nielsen global:** "Green Generation: Millennials Say Sustainability Is a Shopping Priority," Nielsen Global Survey of Corporate Social Responsibility and Sustainability, May 11, 2015, https://www.nielsen.com/eu/en/insights/news/2015/green -generation-millennials-say-sustainability-is-a-shopping-priority.html.

**161 "Millennials want":** Elisa Neimtzow, interview with the author, by Skype, October 12, 2015.

**161 McCartney and her:** David Owen, "Going Solo," *New Yorker*, September 17, 2001, https://www.newyorker.com/magazine/2001/09/17/going-solo.

**161 McCartney was a self-described:** Owen, "Going Solo," 132.

**162 "My mum wore":** "She Hopes You Will Enjoy the Show," *Newsweek*, April 27, 1997.

**162 "Stella Steel":** Owen, 130.

**163 Lagerfeld snipped:** Owen, "Going Solo," 130.

**163 With her first show:** Suzy Menkes, "Glitter-Gulch from Givenchy as McQueen Goes Wild West: A Stellar Start for Chloe's Light-Hearted Little Nothings," *International Herald Tribune*, October 16, 1997, https://www.nytimes.com/1997/10/16 /news/glittergulch-from-givenchy-as-mcqueen-goes-wild-west-a-stellar-start .html.

**163 "What Stella did":** Owen, "Going Solo," 130.

**163 "Livestock production":** Keinan and Crener, "Stella McCartney," 5.

**163 with more than:** Jess Cartner-Morley, "Stella McCartney: 'Fashion People Are Pretty Heartless,'" *Guardian*, October 5, 2009, https://www.theguardian.com/life andstyle/2009/oct/05/stella-mccartney-fashion-heartless.

**163 Conventional leather tanning:** Keinan and Crener, "Stella McCartney," 5.

**164 90 percent:** Carry Hq, "Chrome vs Vegetable Tanned Leather," Carryology, August 28, 2015, https://www.carryology.com/insights/chrome-vs-vegetable-tanned -leather/.

**164 she was swiftly:** Owen, "Going Solo," 132.

**165 one published report:** Imran Amed, "Stella McCartney: A Success Without Making Fashion Victims Out of Animals," *Evening Standard*, April 9, 2015, https://www .standard.co.uk/business/markets/stella-mccartney-a-success-without-making -fashion-victims-out-of-animals-10164406.html.

**166 McCartney's sustainability and ethical:** Claire Bergkamp, interview with the author, London, January 20, 2017. All Bergkamp quotes come from this interview, phone calls, or emails, unless otherwise indicated.

**168 The EP&L "allows":** Keinan and Crener, "Stella McCartney," 16.

**168 In 2014, McCartney:** Betsy Andrews, "What Is 'Rainforest-Free' Clothing, and Why Should You Care?" Racked, March 16, 2017, https://www.racked.com/2017/3/16/14938354/fashion-sustainability-rainforest-free-clothing.

**169 For centuries, cashmere:** Una Jones, interview with the author, by phone, April 26, 2017.

**168 If fashion's demand:** Jones interview.

**169 in 2016, Stella:** "Stella McCartney's Eco-Conscious Solution to Cashmere," *Modem*, August 3, 2016, http://www.modemonline.com/modem-mag/article/3793-united-kingdom—london-stella-mccartneys-eco-conscious-solution-to-cashmere.

**171 Swiss entrepreneur:** Nina Marenzi, interview with the author, London, March 14, 2017. All Marenzi quotes come from this interview.

**172 There was a white:** Luisa Zargani, "Salvatore Ferragamo Launches Capsule Collection Made with Orange Fiber," *Women's Wear Daily*, April 17, 2017, http://wwd.com/fashion-news/designer-luxury/exclusive-salvatore-ferragamo-launches-capsule-collection-made-orange-fiber-10868843/.

**173 economist Robert Reich:** Robert Reich, "Corporations Won't Lead the Way on Solving Global Warming," *American Prospect*, October 18, 2007, https://prospect.org/article/corporations-wont-lead-way-solving-global-warming-0.

**173 at Inditex:** Imran Amed and Kate Abnett, "Inditex: Agile Fashion Force," Business of Fashion, March 30, 2015, https://www.businessoffashion.com/community/voices/discussions/can-fashion-industry-become-sustainable/inditex-agile-fashion-force.

**173 many belong to:** Jason Kibbey, interview with the author, by phone, November 20, 2017.

**174 H&M has pledged:** Nathalie Remy, Eveline Speelman, and Steven Swartz, "Style That's Sustainable: A New Fast-Fashion Formula," McKinsey & Company, October 2016.

**174 "future-proof our business":** Anna Gedda, interview with the author, Copenhagen, May 11, 2017.

**175 While living in Shanghai:** Teresa Novellino, "Modern Meadow Founder Andras Forgacs Makes Leather in a Brooklyn Lab," *New York Business Journal*, October 3, 2016, https://www.bizjournals.com/newyork/news/2016/10/03/modern-meadow-andras-forgacs-reinventor-upstart100.html.

**176 Already, Forgacs:** Eillie Anzilotti, "How Modern Meadow Is Fabricating the Animal-Free Leather of the Future," *Fast Company*, October 11, 2017, https://www.fastcompany.com/40475098/how-modern-meadow-is-fabricating-the-animal-free-leather-of-the-future.

**176 Suzanne Lee, a savvy:** Suzanne Lee, interview with the author, New York, June 21, 2017. All Lee quotes come from this interview unless otherwise indicated.

**176 A year later:** Anzilotti, "How Modern Meadow Is Fabricating the Animal-Free Leather of the Future."

**176 Leather is a $100-billion:** Novellino, "Modern Meadow Founder Andras Forgacs Makes Leather in a Brooklyn Lab."

**176 And consumer demand:** "Global Leather Goods Market 2017-2021," Technavio, September 2017, https://www.technavio.com/report/global-leather-goods-market

?gclid=EAIaIQobChMI9YHxgaqc3gIVIYXVCh1bMA29EAAYASAAEgJELfD
_BwE.

177 **the livestock industry:** Robert Goodland and Jeff Anhang, "Livestock and Climate Change," *World Watch*, November/December 2009, http://www.worldwatch .org/node/6294.

178 **Since my interview:** Jill Meisner, by email, February 26, 2019.

180 **With the help:** Vikram Alexei Kansara, "With Lab-Grown Leather, Modern Meadow Is Engineering a Fashion Revolution," Business of Fashion, September 26, 2017, https://www.businessoffashion.com/articles/fashion-tech/bof-exclusive-with -lab-grown-leather-modern-meadow-is-bio-engineering-a-fashion-revolution.

180 **they hired a trio:** Kansara, "With Lab-Grown Leather."

181 **Spiders extrude:** Amy Feldman, "Clothes from a Petri Dish: $700 Million Bolt Threads May Have Cracked the Code on Spider Silk," *Forbes*, April 15, 2018, https:// www.forbes.com/sites/amyfeldman/2018/08/14/clothes-from-a-petri-dish-700 -million-bolt-threads-may-have-cracked-the-code-on-spider-silk/#4c41535ebda1.

181 **The science behind:** Jamie Bainbridge, interview with the author, Emeryville, CA, October 13, 2017. All Bainbridge quotes are from this interview, phone calls, or emails, unless otherwise indicated.

185 **By mid-2018, Bolt:** Katya Foreman, "This $400 Tote Is Made of Mushrooms, Not Leather," *Women's Wear Daily*, September 8, 2018, https://wwd.com/fashion-news /fashion-features/exclusive-bolt-threads-first-commercialized-mylo -bag-1202783029/.

## CHAPTER EIGHT: AROUND AND AROUND WE GO

188 **"tons of clothing":** Ellen MacArthur and Julie Wainwright, "The New Textiles Economy," Copenhagen Fashion Summit, Copenhagen, May 16, 2018.

188 **To tack in the right direction:** MacArthur and Wainwright, "The New Textiles Economy."

189 **she always visited:** In 2000, the NLC charged that Target sourced from sweatshops in Nicaragua; the company claimed its inspectors saw no abuses, and it has been vigilant in this regard ever since. Carrie Antifinger, "Nicaragua: US Retailers Contract with Sweatshops," Associated Press, August 22, 2000, https://corpwatch.org/article /nicaragua-us-retailers-contract-sweatshops.

189 **"You could eat":** Stacy Flynn, interview with the author, Villepinte, France September 18, 2018. All Flynn quotes come from this interview, or follow-up emails, unless otherwise indicated.

190 **"find solutions that":** Stacy Flynn, speech, Fast Company's World Changing Idea Awards, *Fast Company*, May 7, 2018.

194 **in New York City alone:** Elizabeth Cline, "Where Does Discarded Clothing Go?" *Atlantic*, July 18, 2014, https://www.theatlantic.com/business/archive/2014/07 /where-does-discarded-clothing-go/374613/.

195 **"Wouldn't it be":** Cyndi Rhoades, interview with the author, Eurostar, June 15, 2017. All Rhoades quotes are from this interview or follow-up interviews.

**197 In 2000, he saw Bill:** Craig Cohon, interview with the author, Eurostar, June 15, 2017. All Cohon quotes are from this interview, unless otherwise indicated.

**197 "It was the time":** "In Conversation with Craig Cohon, Live," *Tank*, September 26, 2018, https://tankmagazine.com/tank/2018/09/craig-cohon/.

**197 "I felt he was":** Mark Leonard, "The Coca-Cola Man Who Had a Vision," *New Statesman*, March 11, 2002, https://www.newstatesman.com/node/194377.

**197 partnered with BP:** "In Conversation with Craig Cohon, Live."

**199 In 2016, that equaled:** Rob Walker, "Fashion in New Bid to Be Truly Sustainable," *Guardian*, April 9, 2017, https://www.theguardian.com/fashion/2017/apr/08/fashion-sustainable-clothes-wwf-finland.

**200 Because Aquafil:** Giulio Bonazzi, interview with the author, by phone, November 16, 2018. All Bonazzi quotes come from this interview, unless otherwise indicated.

**201 After four years:** William McDonough, Bert Wouters, and Giulio Bonazzi, "Business Models for a Closed-Loop Fashion System," Copenhagen Fashion Summit, Copenhagen, May 15, 2018.

**202 "You give me":** McDonough, Wouters, and Bonazzi, "Business Models for a Closed-Loop Fashion System."

**202 "The single best":** Rose Marcario, "Repair Is a Radical Act," Patagonia, November 15, 2015, https://www.patagonia.com/worn-wear.html.

**203 Three-fourths of these:** Sharon Edelson, "Experience Matters: A New Eileen Fisher Retail Concept Grows in Brooklyn," *Women's Wear Daily*, August 21, 2018.

**203 Leftovers are fed:** Edelson, "Experience Matters."

**203 "We must share":** Eileen Fisher, "Innovation with the Next Generation," Copenhagen Fashion Summit, Copenhagen, May 11, 2017.

**203 Levi's Chip Bergh:** Roisin O'Connor, "Levi's CEO Explains Why You Should Never Wash Your Jeans," *Independent*, February 18, 2016, https://www.independent.co.uk/life-style/fashion/levis-ceo-explains-why-you-should-never-wash-your-jeans-a6881031.html.

**204 Procter & Gamble's:** McDonough, Wouters, and Bonazzi, "Business Models for a Closed-Loop Fashion System."

**204 "Made with wind":** William McDonough, "Cradle to Cradle, the Circular Economy and the Five Goods," Copenhagen Fashion Summit, Copenhagen, May 11, 2017.

**205 "If you want":** Katrin Ley, interview with the author, Amsterdam, November 12, 2018.

## CHAPTER NINE: RAGE AGAINST THE MACHINE

**208 Ray Kurzweil:** Elizabeth Paton, "Fashion's Future, Printed to Order," *New York Times*, December 5, 2016, https://www.nytimes.com/2016/12/05/business/fashions-future-printed-to-order.html.

**208 Andrew Bolton, the head curator:** Andrew Bolton, *Manus x Machina: Fashion in an Age of Technology* (New York: The Metropolitan Museum of Art, 2016), 19.

**209 Three-dimensional printing:** Sam Rose, "What Was the First 3D Printed Object Created," *FMSblog*, April 27, 2018, https://fmsblog.azurewebsites.net/first-3d-printed-object-created/.

**209 I met "Eeee-reece":** Iris van Herpen, interview with the author, Amsterdam, June 22, 2018. All van Herpen quotes come from this interview unless otherwise indicated.

**212 "I've known about":** Michael Schmidt, interview with the author, Los Angeles, October 9, 2017. All Schmidt quotes come from this interview and follow-up phone calls.

**214 "Actually, it didn't":** Dita Von Teese, interview with the author, by phone, November 19, 2018.

**217 Traditional knitwear manufacturing:** Ben Alun-Jones and Kirsty Watts, interview with the author, London, March 15, 2017. All Alun-Jones and Watts quotes come from this interview or follow-up phone calls.

**222 The "digital consumer":** Vikram Alexei Kansara, "The Sewbots Are Coming!" Business of Fashion, May 16, 2017, https://www.businessoffashion.com/articles /professional/the-sewbots-are-coming.

**223 The team built:** Palaniswamy Rajan, interview with the author, by phone, November 29, 2018. All Rajan quotes come from this interview.

**225 In 2018, SoftWear partnered:** Tara Donaldson, "Li & Fung Enlists Sewbot Technologies in Drive for Digital Supply Chain," Sourcing Journal, May 8, 2018, https:// sourcingjournal.com/topics/technology/li-fung-sewbot-technologies -digital-supply-chain-105527/.

**226 You've got Nike's:** "The Robot Startup Using Static Electricity to Make Nike Sneakers," Business of Fashion, August 30, 3017, https://www.businessoffashion.com/arti cles/fashion-tech/these-robots-are-using-static-electricity-to-make-nike-sneakers.

**226 You've got Adidas's:** Kansara, "The Sewbots Are Coming!"

**226 In 2016, the International Labour:** Jae-He Chang, Gary Rynhart and Phu Huynh, "ASEAN in Transformation: The Future of Jobs at Risk of Automation," International Labour Organization, July 1, 2016, https://www.ilo.org/actemp/publications /WCMS_579554/lang--en/index.htm.

**227 "We'll never do . . .":** Adele Peters, "This T-Shirt Sewing Robot Could Radically Shift the Apparel Industry," *Fast Company*, August 25, 2017, https://www.fastcompany .com/40454692/this-t-shirt-sewing-robot-could-radically-shift-the-apparel-industry.

## CHAPTER TEN: TO BUY OR NOT TO BUY

**232 Moda's return rate is:** Sarah Kennedy, "Exclusive: Lauren Santo Domingo on the Quiet Success of Moda Operandi," *Observer*, January 26, 2016, https://observer.com /2016/01/exclusive-lauren-santo-domingo-on-the-quiet-success-of-moda -operandi/.

**232 "we help a woman":** Lauren Santo Domingo, interview with the author, London, November 18, 2015. All Santo Domingo quotes come from this interview unless otherwise indicated.

**232 Because Moda's full-price:** Kennedy, "Exclusive: Lauren Santo Domingo on the Quiet Success of Moda Operandi." (Moda refused to provide more recent figures.)

**233 "We would sit among:** Kennedy, "Exclusive: Lauren Santo Domingo on the Quiet Success of Moda Operandi."

**233 Bill Blass was the:** Susan Orlean, "King of the Road," *New Yorker*, December 20, 1993, 87-88, https://www.newyorker.com/magazine/1993/12/20/king-of-the-road.

**234 Until that moment:** Robert Burke, interview with the author, by phone, December 12, 2018. All Burke quotes come from this interview.

**235 In 2017, apparel was the number:** "Revenue of Leading E-Retail Categories in the United States in 2017 (in Billion U.S. Dollars)," Statista, n.d., https://www .statista.com/statistics/568830/us-e-retail-sales-by-category/.

**235 Globally, fashion e-commerce:** Arron Orendorff, "The State of the Ecommerce Fashion Industry," Shopify, March 16, 2018, https://www.shopify.com/enterprise /ecommerce-fashion-industry.

**235 Ninety-two percent:** Antonio Achille, Nathalie Remy, and Sophie Marchessou, "The Age of Digital Darwinism," McKinsey & Company, January 2018.

**235 A 2017 study:** Helen Edwards and Dave Edwards, "Why Shoppers Ditch Traditional Stores for Online in Their Twenties," Quartz, November 29, 2017, https://qz.com /1139098/why-shoppers-ditch-traditional-stores-for-online-in-their-twenties/.

**236 In 2017, the Seattle-based:** Lauren Mang, "High-End Stores That Don't Actually Sell Anything Are the Future of Retail," Quartz, November 24, 2017, https://qz.com /1135230/high-end-stores-that-dont-actually-sell-anything-are-the-future-of-retail/.

**238 The socioeconomic theater:** Émile Zola, *The Ladies' Paradise* (Oxford: Oxford University Press, 1998), ix.

**239 In 2017 alone:** Lauren Sherman, "How to Save the Mall," Business of Fashion, October 2, 2017, https://www.businessoffashion.com/articles/intelligence/how-to -save-the-mall.

**240 In early 2019, the struggling:** Michael J. de la Merced and Michael Corkery, "Lord & Taylor Building, Icon of New York Retail, to Become WeWork Headquarters," *New York Times*, October 24, 2017, https://www.nytimes.com/2017/10/24/business /lord-taylor-wework.html. (WeWork has since rebranded to The We Company).

**240 the first three:** David Moin, "WeWork Sets Vision for Lord & Taylor Flagship," *Women's Wear Daily*, October 30, 2018, https://wwd.com/business-news/retail /wework-sets-vision-for-lord-taylor-flagship-1202895358/.

**240 Lord & Taylor's parent:** De la Merced and Corkery, "Lord & Taylor Building, Icon of New York Retail, to Become WeWork Headquarters."

**240 In 2007, Amazon:** Anna Nicolaou, "Now Amazon Is Disrupting Fashion Retail, Too," *Financial Times*, January 26, 2018, https://www.ft.com/content/795935ac -0205-11e8-9650-9c0ad2d7c5b5.

**240 By the end of 2017:** Nicolaou, "Now Amazon Is Disrupting Fashion Retail, Too."

**240 with projected sales:** Daphne Howland, "Amazon Poised to Reign Over Apparel by Years' End," Retail Dive, September 12, 2018, https://www.retaildive.com/news /amazon-poised-to-reign-over-apparel-by-years-end/532151/.

**241 Analysts predict:** Nicolaou, "Now Amazon Is Disrupting Fashion Retail, Too."

**241 possess 16 percent:** Achille, Remy, and Marchessou, "The Age of Digital Darwinism."

**241 Amazon introduced:** Alyssa Pagano, "I Let Amazon's New Echo Look Choose My Clothes for a Week—Here's How It Went," Business Insider, June 6, 2018, https://www.businessinsider.fr/us/amazon-echo-look-alexa-style-assistant-review -2018-5.

**241 Amazon also began:** Lauren Thomas, "Amazon's 100 Million Prime Members Will Help It Become the No. 1 Apparel Retailer in the U.S.," CNBC.com, April 19, 2018, https://www.cnbc.com/2018/04/19/amazon-to-be-the-no-1-apparel-retailer-in-the-us-morgan-stanley.html.

**241 "They're going":** Imran Amed, "Chip Bergh on Steering Levi's Through the Uncertainties of 2017," Business of Fashion, January 8, 2018, https://www.businessoffashion.com/articles/ceo-talk/chip-bergh-on-steering-levis-through-the-uncertainties-of-2017.

**241 To that end, in 2017:** Jason Del Rey, "Amazon Won a Patent for an On-Demand Clothing Manufacturing Warehouse," Recode, April 18, 2017, https://www.recode.net/2017/4/18/15338984/amazon-on-demand-clothing-apparel-manufacturing-patent-warehouse-3d.

**242 Fabric would be:** Marc Bain, "Amazon Has Patented an Automated On-Demand Clothing Factory," Quartz, April 19, 2017, https://qz.com/963381/amazon-amzn-has-patented-an-automated-on-demand-clothing-factory/.

**242 "Imagine if Amazon":** Nicolaou, "Now Amazon Is Disrupting Fashion Retail, Too."

**242 Pop-ups were conjured:** Amanda Fortini, "The Anti-Concept Concept Store," *New York Times Magazine*, December 12, 2004, https://www.nytimes.com/2004/12/12/magazine/anticoncept-concept-store-the.html.

**243 Adidas popped:** Michael Reilly, "Three-D Knitting Brings Tech to Your Sweaters for a Price," *Technology Review*, April 6, 2017, https://www.technologyreview.com/s/604102/3-d-knitting-brings-tech-to-your-sweaters-for-a-price/.

**244 by late 2018, #ootd:** Sheila Marikar, "The Transformational Bliss of Borrowing Your Office Clothes," *New York Times*, October 12, 2018, https://www.nytimes.com/2018/10/12/business/rent-the-runway-office-clothes.html.

**244 what the Amazon UK:** Sarah Butler, "Amazon Opens Pop Up Fashion Shop in Central London," *Guardian*, October 23, 2018, https://www.theguardian.com/technology/2018/oct/23/amazon-opens-pop-up-fashion-shop-in-central-london.

**244 "Every retailer needs":** Samantha Conti, "How Much Faster Can Fashion Get?" *Women's Wear Daily*, April 5, 2017, https://wwd.com/business-news/retail/how-much-faster-fashion-get-10857985/.

**244 Facebook/Instagram global head:** Morin Oluwole, interview with the author, Paris, December 20, 2018.

**246 "When I first presented":** Matthew Bell, "The Selfridges Scion Making a Splash," *Independent*, May 14, 2011, https://www.independent.co.uk/news/business/analysis-and-features/the-selfridges-scion-making-a-splash-2284163.html.

**246 A stylish brunette:** Daniella Vega, interview with the author, London, June 13, 2017. All Vega quotes come from this interview.

**248 The investment and attentiveness:** Elias Jahshan, "Selfridges Reclaims Best Department Store Title," *Retail Gazette*, May 25, 2018, https://www.retailgazette.co.uk/blog/2018/05/selfridges-reclaims-best-department-store-world-title/.

**248 In 2018, the company completed:** "Selfridges Bags Another Record Year Despite Gloomy Retail Outlook," *Irish News*, October 1, 2018, http://www.irishnews

.com/business/2018/10/01/news/selfridges-bags-another-record-year-despite -gloomy-retail-outlook-1447034/.

250 **Wainwright's teams:** Ellen MacArthur and Julie Wainwright, "The New Textiles Economy," Copenhagen Fashion Summit, Copenhagen, May 16, 2018.

251 **Some brands are still:** Jessica Binns, "Resale, Rentals and Subscriptions Have Tommy Hilfiger, Michael Kors Spooked," Sourcing Journal, March 1, 2019, https:// sourcingjournal.com/topics/retail/resale-rentals-subscriptions-tommy-hilfiger -michael-kors-edited-141430/.

251 **In November 2018:** "Chanel Is Suing the RealReal for Allegedly Selling Counter feit Bags," Fashion Law, November 15, 2018, http://www.thefashionlaw.com /home/chanel-is-suing-the-realreal-for-allegedly-selling-counterfeit-bags.

252 **It won't be stopped:** MacArthur and Wainwright, "The New Textiles Economy."

253 **It wasn't easy:** Marikar, "The Transformational Bliss of Borrowing Your Office Clothes."

253 **According to Hyman:** Alexendra Schwartz, "Rent the Runway Wants to Lend You Your Look," *New Yorker*, October 22, 2018.

253 **For $159 a month:** Anna Nicolaou and Mark Vandervelde, "Retailers Respond to Rise in Renting Clothes and Goods," *Financial Times*, December 17, 2017, https:// www.ft.com/content/ca2e1860-e425-11e7-8b99-0191e45377ec.

254 **"The smell has":** Schwartz, "Rent the Runway Wants to Lend You Your Look."

255 **Rent the Runway:** Patricia Marx, "The Borrowers," *New Yorker*, January 31, 2011.

255 **also accumulates a wealth:** Schwartz, "Rent the Runway Wants to Lend You Your Look."

255 **"It's an amazing":** Marikar, "The Transformational Bliss of Borrowing Your Of fice Clothes."

255 **Brochard and Brizay met:** Emmanuelle Brizay, interview with the author, Paris, November 6, 2018. All Brizay quotes come from this interview unless otherwise indicated.

257 **"It's still hush-hush":** Sara Dalloul, interview with the author, Paris, October 26, 2017. All Dalloul quotes come from this interview.

257 **Panoply stylist Bettina:** Bettina Hetoubanabo, interview with the author, Paris, October 26, 2018.

258 **In 2016, Panoply raised:** Harriet Agnew, "Rent-à-Porter—Would You Hire Your Wardrobe?" *Financial Times*, September 28, 2018, https://www.ft.com/content /3211b24c-c171-11e8-95b1-d36dfef1b89a.

258 **To put it in perspective:** Corinne Ruff, "30 Minutes with Rent the Runway's CEO," Retail Dive, May 7, 2018.

258 **In the spring of 2018:** John Mowbray, "France Proposes Law to Tackle Unsold Clothing Problem," *Ecotextile News*, April 25, 2018, https://www.ecotextile.com /2018042523440/fashion-retail-news/france-proposes-law-to-tackle-unsold -clothing-problem.html.

# Selected Bibliography

ANGUELOV, NIKOLAY. *The Dirty Side of the Garment Industry: Fast Fashion and Its Negative Impact on Environment and Society*. Boca Raton, FL: CRC Press, 2016.

BALFOUR-PAUL, JENNY. *Indigo: Egyptian Mummies to Blue Jeans*. Buffalo, NY: Firefly Books, 1998.

BLASS, BILL AND CATHY HORYN, ED. *Bare Blass*. New York: HarperCollins, 2002.

BOLTON, ANDREW. *Manus x Machina: Fashion in an Age of Technology*. New York: The Metropolitan Museum of Art, 2016.

BROWN, JOHN. *A Memoir of Robert Blincoe*. Sussex, UK: Caliban Books, 1977.

CHAZEN, JEROME. *My Life at Liz Claiborne: How We Broke the Rules and Built the Largest Fashion Company in the World*. Bloomington, IN: AuthorHouse, 2011.

DORGAN, BYRON L. *Take This Job and Ship It: How Corporate Greed and Brain-Dead Politics Are Selling Out America*. New York: Thomas Dunne Books, 2006.

DOWNEY, LYNN. *Levi Strauss: The Man Who Gave Blue Jeans to the World*. Amherst: University of Massachusetts Press, 2016.

ENGELS, FRIEDRICH. *The Condition of the Working Class in England*. Oxford: Oxford University Press, 1993.

HARNEY, ALEXANDRA. *The China Price: The True Cost of Chinese Competitive Advantage*. New York: Penguin Press, 2008.

HONORÉ, CARL. *In Praise of Slowness: Challenging the Cult of Speed*. New York: HarperCollins, 2004.

*IRIS VAN HERPEN: Transforming Fashion*. Atlanta: High Museum of Art, 2015.

KOIKE, KAZUKO. *Where Did Issey Come From?* Tokyo: Mizue Nakamura, 2017.

LAVERGNE, MICHAEL. *Fixing Fashion: Rethinking the Way We Make, Market and Buy Our Clothes*. Gabriola Island, BC: New Society, 2015.

LEE, SUZANNE. *Fashioning the Future: Tomorrow's Wardrobe*. London: Thames & Hudson, 2005.

LIVERIS, ANDREW. *Make It in America: The Case for Re-Inventing the Economy*. Hoboken, NJ: John Wiley & Sons, Inc., 2012.

MCCLENDON, EMMA. *Denim: Fashion's Frontier*. New York: Fashion Institute of Technology, 2016.

MCDONOUGH, WILLIAM AND MICHAEL BRAUNGART. *Cradle to Cradle: Remaking the Way We Make Things*. London: Vintage, 2009.

RIVOLI, PIETRA. *The Travels of a T-Shirt in the Global Economy: An Economist Examines the Markets, Power, and Politics of World Trade*. Hoboken, NJ: John Wiley & Sons, Inc., 2015.

ROBINSON, HARRIET H. *Loom and Spindle: Or, Life Among the Early Mill Girls*. Kailua, HI: Press Pacifica, 1976.

SCHOENBERGER, KARL. *Levi's Children: Coming to Terms with Human Rights in the Global Marketplace*. New York: Atlantic Monthly Press, 2000.

SHELL, ELLEN RUPPEL. *Cheap: The High Cost of Discount Culture*. New York: Penguin Press, 2009.

SOYER, DANIEL, ED. *A Coat of Many Colors: Immigration, Globalization, and Reform in New York City's Garment Industry*. New York: Fordham University Press, 2005.

STEIN, LEON, ED. *Out of the Sweatshop: The Struggle for Industrial Democracy*. New York: Quadrangle/New Times Book Company, 1977.

*THE HISTORY OF COTTON*. Virginia Beach: The Donning Company Publishers, 2005.

VON DREHLE, DAVID. *Triangle: The Fire That Changed America*. New York: Grove Press, 2003.

WALDINGER, ROGER D. *Through the Eye of the Needle: Immigrants and Enterprise in New York's Garment Trades*. New York: New York University Press, 1986.

YAFA, STEPHEN. *Cotton: The Biography of a Revolutionary Fiber*. New York: Penguin, 2005.

*YVES SAINT LAURENT*. New York: The Metropolitan Museum of Art, 1983.

# Photo Credits

INTRODUCTION: Melania Trump. © 2018 by Chip Somodevilla/ Getty Images.

CHAPTER ONE: Cate Blanchett at the 71st Cannes Film Festival. © 2018 by George Pimentel/Getty Images.

CHAPTER TWO: Shila Begum at Rana Plaza. © 2018 by Clara Vannucci.

CHAPTER THREE: Sally Fox. © 2012 by Paige Green.

CHAPTER FOUR: Natalie Chanin picking cotton on the Lentz farm in Trinity, Alabama. © 2012 by Rinne Allen.

CHAPTER FIVE: Tower Mill. © 2015 by Chris Bull/Alamy Stock Photo.

CHAPTER SIX: Sarah Bellos in her indigo fields. © 2016 by Larry McCormack/The Tennessean.

CHAPTER SEVEN: Stella McCartney at Bolt Threads. © 2016 by Stephane Jaspar.

CHAPTER EIGHT: Patagonia's Worn Wear Wagon. © 2014 by Erin Feinblatt.

CHAPTER NINE: Iris van Herpen's Anthozoa suit, Spring-Summer Haute Couture 2013 fashion show in Paris. © 2013 by Don Ashby/FirstVIEW.

CHAPTER TEN: Playing "midget" golf on Selfridges's roof garden. © 1930 by General Photographic Agency/Getty Images.

# Index

Kenya, 7

Kering, 164, 166, 167, 168, 173, 180, 199, 204, 250

Kibbey, Jason, 174

Klein, Calvin, 54, 74, 240, 242, 277

Klein, Naomi, 196

knitwear, on-demand, 217–21

Knoxville, Tenn., 81–82

Kodama, Hideo, 209

Kojima, Japan, 142–43

Korten, David C., 196

Kress, François, 180

Kurashiki, Japan, 143

Kurt Salmon Associates, 29, 36, 244

Kurzweil, Ray, 208

labor abuses, 2, 4, 10, 42–43, 47–48, 49–50, 160, 189–90
    *see also* human rights; sweatshops

Labor Department, US, 41, 42, 50–51

La Coruña, Spain, 30, 34

Lacroix, Christian, 162

Lagerfeld, Karl, 162–63

land degradation, 163, 167, 168, 169, 189

landfills, discarded clothing in, 8

Lang, Fritz, 9

Lauren, Ralph, 5

leather:
    consumer demand for, 176–77
    McCartney's ban on, 160, 162, 163–64
    synthetic, 165, 168, 171–72
    *see also* biofabricated material

Le Bon Marché, 238, 243

Lee, Suzanne, 176, 177–78

Lentz, Jimmy, 104

Lentz, Lisa, 104

Lepore, Nanette, 120

Leslie Fay, 47–48, 49

Levin, Sue, 183

Levi Strauss & Co., 72–73, 137
    Blue Ridge plant of, 83–84
    Evrnu and, 191–92
    falling sales at, 80–85
    501 jeans of, 73, 74, 143, 153
    humane business practices of, 80–81, 83–84
    innovation at, 149–53, 207
    Knoxville plant of, 81–82
    manufacturing code of conduct of, 50
    offshoring by, 69, 151
    organic cotton and, 79
    plant closings by, 81–82, 145

Levi's Vintage Clothing (LVC), 153, 154

Ley, Katrin, 205

Li & Fung, 225–26

*Life 3.0* (Tegmark), 210

linen, 172

Little River Sock Mill, 105

livestock, environmental impact of, 163–64, 177

Locklear, Gina, 105

looms, antique, 144–46

Lopez, Dorka Nohemi Diaz, 49

Lord & Taylor, 60, 102, 240

Los Angeles, Calif., 109, 127, 132, 196
    as apparel manufacturing center, 40
    sweatshops in, 39–43, 110

Louis Vuitton, 173

Love, David, 151

"loving the thread," 96–97

Lowell, Francis Cabot, 22–23, 45

Lowell, Josephine Shaw, 45

Lyocell, 183

MacArthur, Ellen, 188–89, 204

McCartney, Linda Eastman, 161

McCartney, Paul, 161

McCartney, Stella, 188, 221, 249, 250–51, 255, 257, 259, 260
    Adidas and, 166
    affordability and, 166
    animal-free ethos of, 160, 162, 163–64, 181
    background and career of, 161–64
    Bolt Threads and, 180–81, 183, 184
    ECONYL and, 168, 201–2, 248
    Evrnu and, 193
    Fall-Winter 2017–2018 women's wear show of, 159–60, 170
    H&M and, 166
    humane business practices of, 170
    PVC ban instituted by, 165–66
    as sustainability advocate, 160–71, 172–73, 175

McCormack, Brendan, 114, 116, 119

McDonough, William, 188, 204

McKinsey researchers, 35, 235

McNeill, Katy, 102, 106

McNeill, K. P., 102, 103–4, 106–7

McQueen, Alexander, 162, 210, 223, 246

madder, 138

"Made in the U.S.A." label, 30, 40–41, 50

Magnúsdóttir, Áslaug, 232

Manchester, England:
    as center of British cotton industry, 9, 22–23, 43–45
    revival of cotton milling in, 113–17

# ABOUT THE AUTHOR

Dana Thomas is the author of *Gods and Kings* and the *New York Times* bestseller *Deluxe*. She began her career writing for the Style section of the *Washington Post*, and she has served as a cultural and fashion correspondent for *Newsweek* in Paris. She is a regular contributor to the *New York Times* Style section and has written for the *New York Times Magazine*, the *New Yorker*, the *Wall Street Journal*, the *Financial Times*, *Vogue*, *Harper's Bazaar*, *T: The New York Times Style Magazine* and *Architectural Digest*. In 2016, the French Minister of Culture named Thomas a Chevalier of the Order of Arts and Letters. She lives in Paris.